David Windisch

Random walks, disconnection and random interlacements

David Windisch

Random walks, disconnection and random interlacements

A dissertation submitted to ETH Zurich

Südwestdeutscher Verlag für Hochschulschriften

Impressum/Imprint (nur für Deutschland/ only for Germany)
Bibliografische Information der Deutschen Nationalbibliothek: Die Deutsche Nationalbibliothek
verzeichnet diese Publikation in der Deutschen Nationalbibliografie; detaillierte bibliografische
Daten sind im Internet über http://dnb.d-nb.de abrufbar.
Alle in diesem Buch genannten Marken und Produktnamen unterliegen warenzeichen-, marken-
oder patentrechtlichem Schutz bzw. sind Warenzeichen oder eingetragene Warenzeichen der
jeweiligen Inhaber. Die Wiedergabe von Marken, Produktnamen, Gebrauchsnamen,
Handelsnamen, Warenbezeichnungen u.s.w. in diesem Werk berechtigt auch ohne besondere
Kennzeichnung nicht zu der Annahme, dass solche Namen im Sinne der Warenzeichen- und
Markenschutzgesetzgebung als frei zu betrachten wären und daher von jedermann benutzt
werden dürften.

Verlag: Südwestdeutscher Verlag für Hochschulschriften Aktiengesellschaft & Co. KG
Dudweiler Landstr. 99, 66123 Saarbrücken, Deutschland
Telefon +49 681 37 20 271-1, Telefax +49 681 37 20 271-0, Email: info@svh-verlag.de
Zugl.: Zürich, ETH Zürich, Dissertation, 2009

Herstellung in Deutschland:
Schaltungsdienst Lange o.H.G., Berlin
Books on Demand GmbH, Norderstedt
Reha GmbH, Saarbrücken
Amazon Distribution GmbH, Leipzig
ISBN: 978-3-8381-1049-3

Imprint (only for USA, GB)
Bibliographic information published by the Deutsche Nationalbibliothek: The Deutsche
Nationalbibliothek lists this publication in the Deutsche Nationalbibliografie; detailed
bibliographic data are available in the Internet at http://dnb.d-nb.de.
Any brand names and product names mentioned in this book are subject to trademark, brand or
patent protection and are trademarks or registered trademarks of their respective holders. The
use of brand names, product names, common names, trade names, product descriptions etc.
even without a particular marking in this works is in no way to be construed to mean that such
names may be regarded as unrestricted in respect of trademark and brand protection legislation
and could thus be used by anyone.

Publisher:
Südwestdeutscher Verlag für Hochschulschriften Aktiengesellschaft & Co. KG
Dudweiler Landstr. 99, 66123 Saarbrücken, Germany
Phone +49 681 37 20 271-1, Fax +49 681 37 20 271-0, Email: info@svh-verlag.de

Copyright © 2009 by the author and Südwestdeutscher Verlag für Hochschulschriften
Aktiengesellschaft & Co. KG and licensors
All rights reserved. Saarbrücken 2009

Printed in the U.S.A.
Printed in the U.K. by (see last page)
ISBN: 978-3-8381-1049-3

Acknowledgments

I extend my sincere appreciation to my adviser Professor Alain-Sol Sznitman for all the time, diligence and patience he has devoted to my benefit. The valuable insights he has shared with me and his pertinent and instructive comments on my work have supported me throughout the completion of this thesis.

Moreover, I am indebted to Professor Erwin Bolthausen, who has kindly agreed to act as the co-examiner.

Abstract

This thesis is concerned with the disconnection of large graphs by trajectories of random walks and the model of random interlacements. The latter is a model for a random set of vertices in a transient graph and has recently been introduced by Sznitman.

Our starting point is the disconnection of a discrete cylinder by a random walk, the most widely studied setup in this area. We consider a random walk on a discrete cylinder of the form $(\mathbb{Z}/N\mathbb{Z})^d \times \mathbb{Z}$, $d \geq 3$, with drift in one of the two $\mathbb{Z}$-directions. The disconnection time is defined as the first time when the trajectory of the random walk disconnects the cylinder into two infinite connected components. We prove that for large N, when the size of the drift exceeds a threshold of $1/N^d$, the disconnection time undergoes a phase transition from polynomial in N to exponential in a power of N.

We then come to a related problem, where we study a random walk trajectory of length uN^d on a discrete torus $(\mathbb{Z}/N\mathbb{Z})^d$ in high dimension d. The set of vertices not visited by the random walk trajectory is referred to as the vacant set. We prove that for parameters $u > 0$ chosen small enough, some components of the vacant set consist only of segments of size logarithmic in N with probability tending to 1 as N tends to infinity. This resolves a question appearing in the work of Benjamini and Sznitman on the giant component of the vacant set.

The remaining part of this work is devoted to the relation between the distributions of the trajectories of the random walks in these two problems and the model of random interlacements.

We first consider the distribution of the configurations of vertices visited by the random walk on $(\mathbb{Z}/N\mathbb{Z})^d$ in the neighborhoods of finitely many distant points in the torus up to time uN^d, $d \geq 3$. This distribution is shown to converge as N tends to infinity to the distribution of independent copies of the random interlacement at level u on $\mathbb{Z}^d$.

Finally, we prove a similar result for random walks on cylinders of the form $G_N \times \mathbb{Z}$, where G_N is a large finite connected weighted graph. We thereby generalize a result proved by Sznitman for G_N the $d \geq 2$-dimensional integer torus of large side length N to a large class of weighted graphs G_N including large Euclidean boxes with reflecting boundary, Sierpinski graphs and trees.

Kurzfassung

Diese Dissertation befasst sich mit der Trennung grosser Graphen durch Irrfahrttrajektorien und dem Random Interlacement Modell. Dieses Modell beschreibt zufällige Teilmengen von Knoten in transienten Graphen und wurde in einer neulich erschienenen Arbeit von Sznitman definiert.

Wir beginnen mit einer mittlerweile oft betrachteten Problemstellung, der Trennung eines diskreten Zylinders durch eine Irrfahrt. Hier betrachten wir eine Irrfahrt auf einem diskreten Zylinder der Form $(\mathbb{Z}/N\mathbb{Z})^d \times \mathbb{Z}$, $d \geq 3$, mit einseitiger Neigung in eine der beiden $\mathbb{Z}$-Richtungen. Die Trennzeit ist definiert als die erste Zeit, zu der die Trajektorie der Irrfahrt den Zylinder in zwei unendliche zusammenhängende Komponenten trennt. Wir zeigen, dass das asymptotische Verhalten der Trennzeit für $N \to \infty$ einen sogenannten Phasenübergang aufweist: Sobald die Neigung der Irrfahrt die Grössenordnung $1/N^d$ überschreitet, ändert sich das Verhalten der Trennzeit von polynomial in N zu exponential in einer Potenz von N.

Danach kommen wir zu einem verwandten Problem, bei dem wir eine Irrfahrttrajektorie der Länge uN^d auf einem diskreten Torus der Seitenlänge N in hoher Dimension d betrachten. Die Menge der von der Irrfahrt nicht besuchten Knoten wird als die vakante Menge bezeichnet. Wir zeigen, dass für ausreichend klein gewählte Parameter $u > 0$ Komponenten der vakanten Menge existieren, die nur aus einem Segment logarithmischer Länge in N bestehen, mit gegen 1 strebender Wahrscheinlichkeit für $N \to \infty$. Dieses Resultat beantwortet eine Frage, die in der Arbeit von Benjamini und Sznitman über die grösste Komponente der vakanten Menge auftritt.

Der übrige Teil dieser Arbeit ist dem Zusammenhang zwischen den Wahrscheinlichkeitsverteilungen der Irrfahrttrajektorien in den obigen Problemstellungen und dem Random Interlacement Modell gewidmet.

Zunächst betrachten wir die Verteilung der Konfigurationen von Knoten, die von der Irrfahrt auf $(\mathbb{Z}/N\mathbb{Z})^d$ in den Umgebungen endlich vieler entfernter Punkte im Torus bis zur Zeit uN^d besucht werden, wobei $d \geq 3$ ist. Wir zeigen, dass diese Verteilung für $N \to \infty$ gegen die Verteilung unabhängiger Kopien eines Random Interlacements auf $\mathbb{Z}^d$ mit Niveau u konvergiert.

Schliesslich zeigen wir ein ähnliches Resultat für Irrfahrten auf Zylindern der Form $G_N \times \mathbb{Z}$, wobei G_N einen grossen endlichen zusammenhängenden gewichteten Graphen bezeichnet. Dieses Resultat wurde von Sznitman für den Fall, in dem G_N der $d \geq 2$-dimensionale diskrete Torus der Seitenlänge N ist, bewiesen und wird hiermit für eine breite Klasse von Graphen G_N verallgemeinert. Diese Klasse enthält unter anderem grosse euklidische Gitter mit reflektierendem Rand, Sierpinski-Graphen und Bäume.

Contents

Acknowledgments   1

Abstract   3

Kurzfassung   5

Introduction   9
1. Random walk trajectories - a brief survey   9
2. Results   14
3. Organization of the thesis   23

Chapter 1. Disconnection of a discrete cylinder by a biased random walk   25
1. Introduction   25
2. Definitions, notation and a useful estimate   31
3. Upper bound   34
4. Lower bounds: Reduction to large deviations   42
5. More geometric lemmas   46
6. The large deviation estimate   55

Chapter 2. Logarithmic components of the vacant set for random walk on a discrete torus   69
1. Introduction   69
2. Some definitions and useful results   72
3. Profusion of logarithmic components until time a_1 74
4. Survival of a logarithmic segment   82
5. Proof of the main result   86

Chapter 3. Random walk on a discrete torus and random interlacements   89
1. Introduction   89
2. Preliminaries   92
3. Proof   95

Chapter 4. Random walks on discrete cylinders and random interlacements   103
1. Introduction   103
2. Notation and hypotheses   109
3. Auxiliary results on excursions and local times   117
4. Excursions are almost independent   122

5.	Proof of the result in continuous time	126
6.	Estimates on the jump process	137
7.	Proof of the result in discrete time	143
8.	Examples	144

Bibliography   159

Introduction

A random walk is a mathematical model describing the erratic movement of a particle. Random walks have wide applicability in many fields of science and engineering. Within mathematics, random walks are intimately linked with discrete potential theory. Before describing the results of the present work in Section 2 of this introduction, we give a brief outline of some related work and explain the title of this thesis in the first section.

1. Random walk trajectories - a brief survey

1.1. Random walks and classical covering problems. This thesis is concerned with trajectories of random walks on graphs. A graph is a set of vertices, some of which are linked by edges. Two vertices linked by an edge are called neighbors. An example of a graph is the two-dimensional square grid, depicted in Figure 1, p. 10, where every vertex has four neighbors. A particle performing a random walk on the vertices of the graph is governed by the following rule: at every step, the particle moves from the current vertex to a neighbor chosen uniformly at random, where each choice is independent of all previous choices. The trajectory of the random walk until time n is the set of vertices visited by the random walk in its first n steps. A realization of a random walk trajectory on a square grid can be seen in Figure 1. Questions about this random set are generally simple to formulate, yet intriguingly difficult to answer. Here are some classical examples.

The setup that has attracted most attention is the random walk on the integer lattice $\mathbb{Z}^d$, $d \geq 1$ (for $d = 2$, this is the example shown in Figure 1). A nontrivial question one can ask in this context is whether every vertex in $\mathbb{Z}^d$ will eventually be visited by the random walk. This question has been answered in a classical work by Pólya in 1921. In [24], Pólya shows that the answer depends on the dimension d: with probability 1, the random walk visits all vertices of $\mathbb{Z}^d$ in dimensions $d = 1$ and $d = 2$, whereas some vertices are never visited in dimensions $d \geq 3$.

Erdős and Révész (see [26] and [16]) have studied the radius of the largest Euclidean ball covered by the random walk trajectory on $\mathbb{Z}^d$ until time n. Their results show that in dimension $d \geq 3$, this radius behaves roughly like $(\log n)^{1/(d-2)}$ for large n with probability 1. In

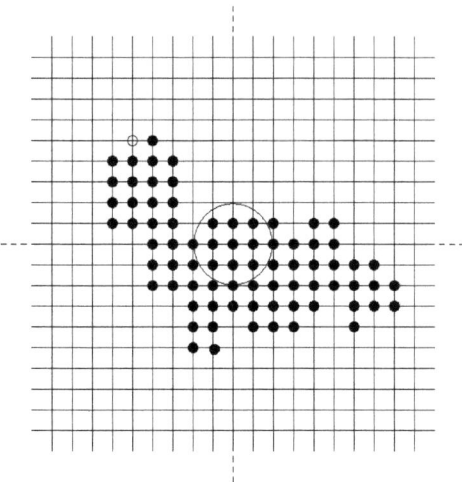

Figure 1: A random walk trajectory on a square grid. The marked vertices have been visited by the random walk starting at the center at time 0 in its first 250 steps and form the random walk trajectory until time 250. The random walk at time 250 is positioned at the vertex marked o. The circle marks the circumference of the largest covered open disc centered at the starting point.

$d = 2$, the problem is substantially more delicate. It took considerable effort (see, for example, Révész [27], Lawler [20]) until Dembo, Peres and Rosen showed in 2007 (see [11]) that the radius of the largest covered disk roughly behaves like $n^{1/4}$ in two dimensions. This behavior differs radically from that of the largest covered disk centered at the starting point of the random walk. Precise asymptotics of order $\exp((const.)(\log n)^{1/2})$ for the radius of this disk were obtained by Dembo, Peres, Rosen and Zeitouni in [10]. In Figure 1, the radius of the largest covered open disc centered at the starting point equals 2.

On finite graphs, one can ask how many steps it takes until the random walk trajectory has covered every vertex. This random time is commonly referred to as the cover time. The most prominent example is perhaps the integer torus $(\mathbb{Z}/N\mathbb{Z})^d$ with large side length N (equipped with edges between any two vertices at Euclidean distance 1). For dimensions $d \geq 3$, the cover time of the torus behaves like $c_d N^d \log N^d$ for large N with high probability, where c_d denotes the expected number of times the random walk on $\mathbb{Z}^d$ visits its starting

1. Random walk trajectories - a brief survey   11

point (see Aldous and Fill [**3**], Chapter 7, p. 22). Although it has been
known for some time that in dimension $d = 2$, the cover time behaves
roughly like $(N \log N)^2$, it is a fairly recent result due to Dembo, Peres,
Rosen and Zeitouni [**10**] which proves that the precise asymptotics for
the cover time in two dimensions are given by $(4/\pi)(N \log N)^2$.

Figure 2: A computer simulation of the largest component (gray)
and second largest component (black) of the vacant set
left by a random walk on $(\mathbb{Z}/N\mathbb{Z})^3$ after $[uN^3]$ steps, for
$N = 200$. The picture on the left-hand side corresponds
to $u = 1$, the right-hand side to $u = 3.5$.

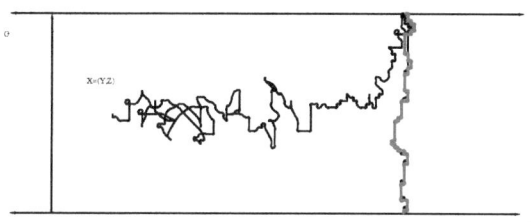

Figure 3: The disconnection time T_N of $(\mathbb{Z}/N\mathbb{Z})^d \times \mathbb{Z}$ by a random walk $(X_n)_{n \geq 0}$ is defined as the first time when the
vacant set has two infinite connected components, separated from each other by some interface covered by the
random walk.

1.2. Disconnection and random interlacements. The present
work is concerned with a phenomenon called disconnection. Let us give
a heuristic visualization of this phenomenon on an example. We again

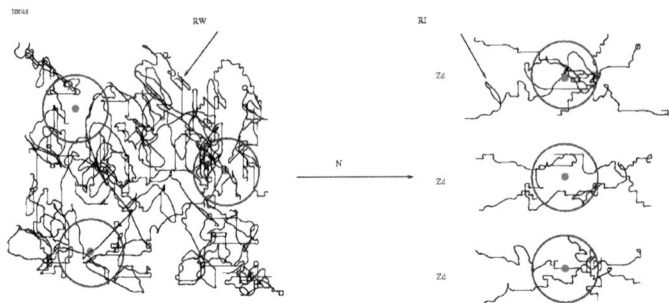

Figure 4: In Chapter 3, we prove that the random configurations of vertices visited by a random walk on $(\mathbb{Z}/N\mathbb{Z})^d$ until time $[uN^d]$ in a fixed number of distant neighborhoods converge in distribution as $N \to \infty$ to the random configurations one obtains from independent copies of the random interlacement on $\mathbb{Z}^d$ at level $u \geq 0$ ($d \geq 3$).

Figure 5: A computer simulation of the largest component (gray) and second largest component (black) of the vacant set of a random interlacement at level u on $\mathbb{Z}^3$, intersected with $[0, N]^3$, for $N = 200$. The picture on the left-hand side corresponds to $u = 1$, the right-hand side to $u = 3.5$. These pictures are strikingly similar to the ones obtained for random walk on $(\mathbb{Z}/N\mathbb{Z})^3$, shown in Figure 2.

look at a random walk on the integer torus and now consider the behavior of the set of vertices not belonging to the random walk trajectory. This set is referred to as the vacant set. The vacant set consists of only

1. Random walk trajectories - a brief survey

one large component (and other components of small size) if we let the random walk run for a small number of steps. This regime is illustrated on the left-hand side of Figure 2, where the second largest component (in black) is still much smaller than the largest component (in gray) of the vacant set. Once the number of time steps of the random walk exceeds some threshold time, this large component splits into many small components (cf. the right-hand side of Figure 2). This phenomenon is what we heuristically mean by disconnection and the time at which it occurs is typically known as the disconnection time.

Disconnection was first investigated on a discrete cylinder $(\mathbb{Z}/N\mathbb{Z})^d \times \mathbb{Z}$, $d \geq 1$, in a seminal work by Dembo and Sznitman [**12**]. In this context, the natural definition of the disconnection time T_N is the first time when the vacant set has two infinite components, see Figure 3, p. 11. In [**12**], the authors prove that the disconnection time behaves roughly like N^{2d} with probability tending to 1 for large N. In Chapter 1 (see also Section 2.1 below), we study the same model for a random walk with bias Δ in the $\mathbb{Z}$-direction and prove that for $d \geq 3$, the disconnection time undergoes a phase transition at $\Delta = 1/N^d$: as long as $\Delta < 1/N^d$, T_N still roughly behaves like N^{2d} for large N, whereas for $\Delta > 1/N^d$, the asymptotic behavior of T_N becomes exponential in a power of N.

Observations on the disconnection of a discrete integer torus

$$(\mathbb{Z}/N\mathbb{Z})^d$$

by a random walk for $d \geq 3$ are up to now largely based on computer simulations. These indicate that as N tends to infinity, disconnection takes place at times of order N^d, significantly smaller than the order $N^d \log N$ of the cover time. More specifically, if the random walk on the torus runs for a time of uN^d for a fixed positive parameter u, then for small u, there is only one large connected component of the vacant set occupying a large fraction of the total volume, whereas for large u, the vacant set typically consists of small components only. For a typical realization of these two regimes, we refer again to Figure 2. Some of these observations are proved by Benjamini and Sznitman who in [**6**] show the existence of a well-defined giant component of the vacant set with high probability in the small u regime for large dimensions d. This investigation is continued in Chapter 2 of this thesis (see also Section 2.2 below), where we show that in the same regime, one typically finds segments of length logarithmic in N among the small components of the vacant set.

Finally, let us mention the model of random interlacements. A random interlacement is a random subset of vertices of the integer lattice $\mathbb{Z}^d$, $d \geq 3$. Random interlacements have been introduced by Sznitman in [**35**] with the aim of describing the microscopic structure of the random walk trajectories in the last two problems. The random interlacement at level $u \geq 0$ consists of a countable collection of doubly infinite random paths on $\mathbb{Z}^d$, where the parameter u governs the amount of trajectories that are present. In Chapter 3 (see also Section 2.3 below), we prove that the random walk trajectory on the integer torus at time uN^d is indeed close in distribution to a random interlacement at level u in microscopic neighborhoods. Figure 4, p. 12, is an attempt to illustrate this result. Given this result, one would expect to be able to advance the understanding of disconnection of the torus by studying properties of the random interlacement. A crucial role should be played by percolative properties of the vacant sets, i.e. of the complements, of the random interlacements, investigated in [35], [30] and [40]. If one compares the vacant set left by random walk on $(\mathbb{Z}/N\mathbb{Z})^3$ after $[uN^3]$ steps with the vacant set of a random interlacement at level u on $\mathbb{Z}^3$ intersected with a box of side length N in computer simulations, the pictures one obtains are very similar indeed (see Figure 5, p. 12).

In the same way, random interlacements can be linked to random walk trajectories on discrete cylinders. A link at the level of microscopic neighborhoods has been established by Sznitman in [36] and is generalized to a larger class of cylinders in Chapter 4 of this thesis, see also Section 2.4 below. In this context, random interlacements have already proved useful for substantial improvements of both upper and lower bounds on the disconnection time of a discrete cylinder, see the works of Sznitman [37] and [38].

2. Results

We now proceed to a more precise description of the results of this thesis.

2.1. Disconnection of a discrete cylinder by a biased random walk. In Chapter 1, we consider a discrete cylinder

$$E = (\mathbb{Z}/N\mathbb{Z})^d \times \mathbb{Z},$$

viewed as a graph with an edge between any two vertices at Euclidean distance 1, for $d \geq 3$ (explanations of this choice of d follow after Theorem 1). Recall that in this context, the disconnection time T_N of E by a process $(X_n)_{n \geq 0}$ on E is defined as the first time when the set $E \setminus \{X_1, \ldots, X_{T_N}\}$ contains two distinct infinite connected components, see Figure 3. Chapter 1 is devoted to the investigation of T_N when $(X_n)_{n \geq 0}$ is a random walk on E with a bias in the $\mathbb{Z}$-direction.

One of the results of Dembo and Sznitman in [12] asserts that if $(X_n)_{n \geq 0}$ is the simple random walk on E, the disconnection time behaves like $N^{2d+o(1)}$ as N tends to infinity. In the present context, we perturb the transition probabilities of the simple random walk with

2. Results

a drift of size $N^{-d\alpha}$ in the positive $\mathbb{Z}$-direction for $\alpha > 0$ and hence obtain a biased random walk. Our main result in Chapter 1 shows that in this setup, the asymptotic behavior of the disconnection time exhibits a phase transition for $\alpha = 1$ in the following sense: the large N behavior of T_N is the same as in the case without drift considered in [12] as long as $\alpha > 1$, and becomes exponential in a power of N when $\alpha < 1$. Here is the statement of the main theorem:

THEOREM 1. (*Theorem 1.1, Chapter 1, p. 26*) *For $d \geq 3$ and any $\epsilon > 0$,*

$$N^{2d-\epsilon} \leq T_N \leq N^{2d+\epsilon}, \quad \text{for } \alpha > 1,$$
$$\exp\left(N^{d(1-\alpha-\varphi(\alpha))-\epsilon}\right) \leq T_N \leq \exp\left(N^{d(1-\alpha)+\epsilon}\right), \quad \text{for } 0 \leq \alpha < 1,$$

with probability tending to 1 as $N \to \infty$, for an explicit piecewise linear function $\varphi : (0,1) \to \left(0, \frac{1}{d-1}\right)$ satisfying $\lim_{\alpha \to 0} \varphi(\alpha) = \lim_{\alpha \to 1} \varphi(\alpha) = 0$ (see Figure 6).

In addition to this result, we reduce the lower bound on T_N to a large deviations problem on the disconnection of a finite box by an unbiased random walk. This problem has some similarity to issues encountered in the study of the vacant set for random walk on the integer torus by Benjamini and Sznitman [6]. The derivation of this large deviation estimate is the point where we use the assumption $d \geq 3$ (the upper bound on T_N is in fact proved for $d \geq 2$). More details on this problem will be mentioned below (see (2.2)).

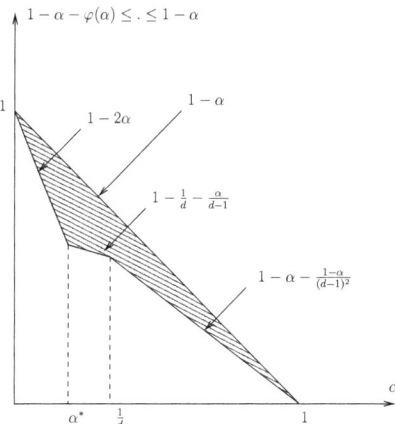

Figure 6: The shaded region lies between the exponents of the upper and lower bounds in Theorem 1 for $\alpha \in (0,1)$.

The upper bound. The derivation of the upper bounds on T_N is based on the observation that the cylinder E is certainly disconnected when a slice of the form $(\mathbb{Z}/N\mathbb{Z})^d \times \{z\} \subseteq E$ is completely covered by the random walk. To exploit this observation, we record visits made to a fixed slice by the random walk. The process of the successive visits to $(\mathbb{Z}/N\mathbb{Z})^d \times \{z\}$ forms a Markov chain on $(\mathbb{Z}/N\mathbb{Z})^d \times \{z\}$ for which we can show similar mixing properties as for the simple random walk on the integer torus. By a coupon-collector estimate on the cover time, this implies that the slice is covered with overwhelming probability after $N^{d+\epsilon}$ visits. It only remains to find an upper bound on the time until $N^{d+\epsilon}$ visits to $(\mathbb{Z}/N\mathbb{Z})^d \times \{z\}$ occur for some $z \in \mathbb{Z}$. This is a simple one-dimensional problem.

The lower bound. The derivation of the lower bounds is significantly more delicate. We reduce the problem of finding a lower bound on T_N to a large deviations problem which could be expressed in words as "How costly is the disconnection of a box by only a few excursions?" and is illustrated in Figure 7. The set whose disconnection concerns us in this problem is the box

$$(2.1) \quad B(\alpha) = \left[-\left[\frac{N}{4}\right], \left[\frac{N}{4}\right]\right]^d \times \left[-\left[\frac{N^{d\alpha \wedge 1}}{4}\right], \left[\frac{N^{d\alpha \wedge 1}}{4}\right]\right] \subset E.$$

We define U_α as the first time when the trajectory of the unbiased simple random walk on E has disconnected the set $B(\alpha)$, in the sense that the random walk has covered some interface carving out two distinct subsets of $B(\alpha)$ of volume at least $|B(\alpha)|/3$. Let D_k be the time at which the random walk completes the k-th excursion between the box $(\mathbb{Z}/N\mathbb{Z})^d \times [-2N^{d\alpha\wedge 1}, 2N^{d\alpha\wedge 1}] \supset B(\alpha)$ and the complement of the larger box $(\mathbb{Z}/N\mathbb{Z})^d \times [-4N^{d\alpha\wedge 1}, 4N^{d\alpha\wedge 1}]$. Then we require an exponential upper bound on the event $U_\alpha \leq D_{[N^\beta]}$, illustrated in Figure 7. In other words, the problem is to find a large deviation estimate for the unbiased random walk of the form

$$(2.2) \qquad P[U_\alpha \leq D_{[N^\beta]}] \leq \exp(-N^{\xi(\alpha,\beta)}).$$

The reduction of the derivation of the lower bounds on T_N to a problem of this form proceeds by a purely geometric argument in the spirit of Dembo and Sznitman, showing that at time T_N, a set of the form $x + B(\alpha) \subset E$ must have been disconnected in the above sense. A Girsanov-type control shows that the drift does not play a role at the scale $N^{d\alpha\wedge 1}$, so that it is sufficient to have an estimate of the form (2.2) for an unbiased random walk. This method yields an explicit relation between the exponent $\xi(\alpha,\beta)$ in the above problem and the exponents appearing in the lower bounds on the disconnection time.

There is a chance that the probability in (2.2) decays only if $\beta \leq d - d\alpha \wedge 1$, because after more than $N^{d-d\alpha\wedge 1+\epsilon}$ excursions, the random walk has typically spent a total time of more than $N^{d+d\alpha\wedge 1+\epsilon}$ in the box

2. Results

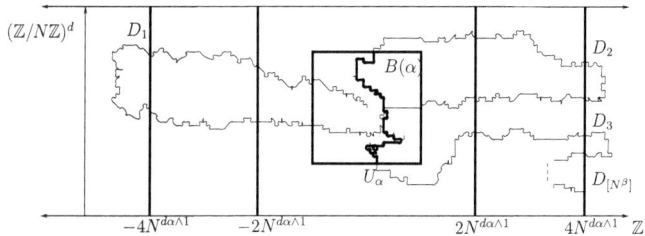

Figure 7: In the large deviation problem (2.2), we need an upper bound on the probability that the first $[N^\beta]$ excursions in and out of the two concentric boxes of length $4N^{d\alpha \wedge 1}$ and $8N^{d\alpha \wedge 1}$ separate two components of the box $B(\alpha)$ of volume $|B(\alpha)|/3$ from each other.

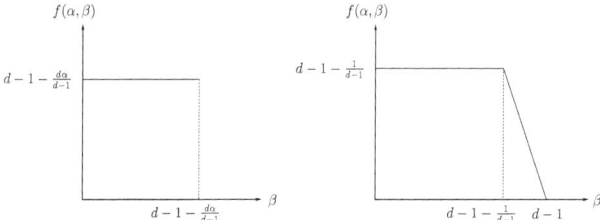

Figure 8: The estimate (2.2) is derived for all $\xi < f$, where f is the function shown above (for $\alpha \in \left(0, \frac{1}{d}\right)$ on the left, for $\alpha \in \left[\frac{1}{d}, \infty\right)$ on the right).

$B(\alpha)$, thereby visited all its vertices and hence already performed the disconnection. If one could prove the bound (2.2) for all such β and all $\xi(\alpha, \beta) < d - d\alpha \wedge 1$, then the lower bound on the disconnection time T_N would match the upper bound for $\alpha < 1$ in Theorem 1. We can prove the estimate (2.2) for all $\xi < f$, with f shown in Figure 8. This derivation requires substantial refinement at the geometric level. We show that at time U_α, one can find many small cubes of side length N^γ, $0 < \gamma < d\alpha \wedge 1$, spread throughout $B(\alpha)$, such that each one of these cubes contains at least $N^{d\gamma}$ vertices visited by the random walk. Note that the number $N^{\gamma d}$ of vertices visited in each $(d+1)$-dimensional cube of side length N^γ corresponds to the number of points required to form an interface separating two macroscopically large components of the cube from each other. A key ingredient for the proof of this geometric result is the application of an isoperimetric inequality. We note here that the restriction of $N^{d\gamma}$ on the minimal number of visited points is significant only when $d \geq 3$. Indeed, the typical number of points visited by the simple random walk at each visit to one of the small cubes of side length N^γ is of order $N^{2\gamma}$, regardless of their dimension $d+1$. This is the reason why we need the assumption $d \geq 3$ in these large deviation estimates. We then prove a tail estimate on the number of points in the small cubes visited by the random walk before exiting a large set. This yields an estimate of the form (2.2) and hence lower bounds on the disconnection time by the previous reduction step.

2.2. Vacant set for random walk on a discrete torus. In Chapter 2, we consider the d-dimensional integer torus $(\mathbb{Z}/N\mathbb{Z})^d$ for a sufficiently large dimension d, equipped with the usual graph structure. We investigate properties of the vacant set, defined as the set of points not visited by the simple random walk $(X_n)_{n\geq 0}$ on $(\mathbb{Z}/N\mathbb{Z})^d$ after $[uN^d]$ time steps for a small fixed parameter $u > 0$ and large N. Here one might wonder why this choice of time steps is interesting. There are at least two answers to this question. First, as we have mentioned in Section 1.2, computer simulations strongly indicate that disconnection of the vacant set into small components occurs at times of this order, see also Figure 2. And second, better understanding of the vacant set can lead to better understanding of the disconnection time of a discrete cylinder mentioned in the last section. Indeed, the behavior of the random walk trajectory at times of this order mimics the configuration of sites visited by the random walk on the discrete cylinder at the time of disconnection in a typical box of the form $(\mathbb{Z}/N\mathbb{Z})^{d-1} \times [z, z+N]$. This observation is in particular applied by Dembo and Sznitman [13] in order to improve the lower bound on the disconnection time of the discrete cylinder.

The investigation of the vacant set was initiated by Benjamini and Sznitman [6]. Their work shows that when d is large, there typically exists a well-defined giant component of the vacant set for small parameters $u > 0$. The definition of the giant component proceeds via segments. Segments are subsets of the torus of the form $\{x, x+e_i, \ldots, x+le_i\}$, where $(e_i)_{1\leq i \leq d}$ denotes the canonical basis of $\mathbb{R}^d$. It is shown in [6] that for a suitably defined constant $c_0 > 0$ depending only on d, many segments of length $l \geq [c_0 \log N]$ remain vacant with probability tending to 1 as N tends to infinity. The authors of [6] then prove that there typically exists a unique component of the vacant set including all vacant segments of length $l \geq [c_0 \log N]$, provided $u > 0$ is chosen small enough. This component is referred to as the giant component. Benjamini and Sznitman prove further that for large N, the giant component is typically at $|.|_\infty$-distance of at most N^β from any point and occupies at least a constant fraction γ of the total volume of the torus for arbitrary $\beta, \gamma \in (0,1)$, when $u > 0$ is chosen sufficiently small. One of the many natural questions that arise

2. Results

is whether the giant component is the only component including segments of a length logarithmic in N. In Chapter 2, we show that the answer is no:

THEOREM 2. (*Theorem 1.1, Chapter 2, p. 70*) *For large dimensions d, there is a constant $c_1 \in (0, c_0)$ depending only on d, such that the vacant set left by the random walk on $(\mathbb{Z}/N\mathbb{Z})^d$ up to time $[uN^d]$ includes some segment of length $[c_1 \log N]$ that does not belong to the giant component with probability tending to 1 as N tends to infinity, provided $u > 0$ is chosen small enough.*

We prove this result by showing that for small $u > 0$, there exists some component of unvisited sites consisting only of a single segment of length $[c_1 \log N]$ with overwhelming probability. The fact that this event is not monotone in the number of random walk steps a priori obstructs the use of a renewal technique. Our approach is to split up the proof into the following two statements:

(1) The set $(\mathbb{Z}/N\mathbb{Z})^d \setminus \{X_0, \ldots, X_{[N^{2-1/10}]}\}$ has at least $[N^\nu]$ components consisting only of a segment of length $[c_1 \log N]$ for some $\nu > 0$ with high probability.
(2) At least one of these $[N^\nu]$ small components is still not visited by the random walk until time $[uN^d]$ with high probability.

For the first statement, we use that the random walk on the torus essentially behaves like simple random walk on $\mathbb{Z}^d$ up to times smaller than N^2. This observation allows us to use non self-intersection properties of the random walk trajectory in high dimension, together with a renewal technique.

For the second statement, we first consider the event A_x that one fixed segment of length $[c_1 \log N]$ with endpoint $x \in (\mathbb{Z}/N\mathbb{Z})^d$ remains unvisited for a time of at least $[uN^d]$. We find a lower bound of of $1/N^\epsilon$, $\epsilon \ll \nu$, on the probability of any such event A_x. It remains to show that the mean $N^{\nu-\epsilon}$ of the number of unvisited segments dominates its standard deviation. To this end, we employ a technique from [6] to bound the covariance of events A_x and $A_{x'}$ for sites x and x' with large distance.

2.3. Random walk on a discrete torus and random interlacements. In Chapter 3, we consider the same setup as in the last section for dimensions $d \geq 3$ and study the connections between the microscopic structure of the set of points visited by the simple random walk on the integer torus and the model of random interlacements on $\mathbb{Z}^d$, $d \geq 3$. Let us briefly motivate this investigation.

As we have mentioned, the chief goal in the study of the vacant set left by a random walk on $(\mathbb{Z}/N\mathbb{Z})^d$ in high dimension d is to prove that the vacant set has only one macroscopically large component for small values of $u > 0$ and only microscopic components for large values of u. An analogous question can be formulated and has been successfully answered for random interlacements. Recall that the random interlacement consists of a countable collection of doubly infinite random paths on $\mathbb{Z}^d$, $d \geq 3$, and that a positive parameter u measures the amount of trajectories that are present. A natural question is whether the vacant set (i.e. the set of vertices of $\mathbb{Z}^d$ not visited by any path) has an infinite component, in other words whether the vacant set percolates. Sznitman [35] and Sidoravicius and Sznitman [30] prove the existence of a positive and finite critical parameter u_* such that with probability one, the vacant set does percolate for $u < u_*$ and does not percolate for $u > u_*$. If one can relate the random configurations given by the set of sites visited by a random walk on the torus until time uN^d to the random interlacement at level $u \geq 0$, then these questions and results on percolation may lead to a better understanding of the disconnection phenomena. In particular, one would expect the parameter u_* to play a role in the transition between the two different regimes for random walk on the torus.

Random interlacements were constructed by Sznitman in [39] and have subsequently been studied in several works (see [40], [41], [30], [31]). The random interlacement at level $u \geq 0$ is the trace left on $\mathbb{Z}^d$ by a cloud of paths constituting a Poisson point process on the space of doubly infinite trajectories modulo time-shift, tending to infinity at positive and negative infinite times. The parameter u is a multiplicative factor of the intensity measure of this point process. It is shown in [35] that the random interlacement is an infinite connected random subset of $\mathbb{Z}^d$, ergodic under translation. Its complement is referred to as the vacant set at level u. The law on $\{0,1\}^{\mathbb{Z}^d}$ of the indicator function of the vacant set at level $u \geq 0$ is denoted by $\mathbb{Q}_u$. An important property of $\mathbb{Q}_u$ is the following characterization:

$$(2.3) \qquad \mathbb{Q}_u[\omega(x) = 1, \text{ for all } x \in A] = \exp\bigl(-u \operatorname{cap}(A)\bigr),$$

for all finite sets $A \subseteq \mathbb{Z}^d$, where we have used the following notation: by $\omega(x)$, $x \in \mathbb{Z}^d$ we denote the canonical coordinates on $\{0,1\}^{\mathbb{Z}^d}$ and by $\operatorname{cap}(A)$ the capacity of A. The capacity is defined as the sum over all points x in A of the probabilities that the simple random walk started at x escapes from A, in other words:

$$\operatorname{cap}(A) = \sum_{x \in A} P_x^{\mathbb{Z}^d}[\{S_1, S_2, \ldots\} \cap A = \emptyset],$$

with $P_x^{\mathbb{Z}^d}$ denoting the law under which the process $(S_n)_{n \geq 0}$ is a simple random walk on $\mathbb{Z}^d$ starting at x.

2. Results

Given $x \in (\mathbb{Z}/N\mathbb{Z})^d$, the vacant configuration left by the walk in the neighborhood of x at time $t \geq 0$ is the $\{0,1\}^{\mathbb{Z}^d}$-valued random variable

(2.4) $\qquad \omega_{x,t}(.) = 1\{X_m \neq \pi(.) + x, \text{ for all } 0 \leq m \leq [t]\}$,

where π denotes the canonical projection from $\mathbb{Z}^d$ onto $(\mathbb{Z}/N\mathbb{Z})^d$ and $(X_n)_{n \geq 0}$ the simple random walk on $(\mathbb{Z}/N\mathbb{Z})^d$. In Chapter 2, we show that vacant sets of random interlacements are the limiting objects of these microscopic pictures in the following sense:

THEOREM 3. (*Theorem 1.1, Chapter 3, p. 90*) *For any M sequences of points $x_1, \ldots, x_M$ in $(\mathbb{Z}/N\mathbb{Z})^d$, $d \geq 3$, of diverging mutual distance,*

$$(\omega_{x_1, uN^d}, \ldots, \omega_{x_M, uN^d}) \text{ converges in distribution to } \mathbb{Q}_u^{\otimes M},$$

as N tends to infinity.

In the proof of Theorem 3, we approximate the distribution of the first time when the random walk on the torus visits a small subset of sites by an exponential distribution, with the help of a result by Aldous and Brown [2]. We then exploit the analogy between random walks on graphs and electric networks, in particular the characterization of the capacity appearing in (2.3) given by the Dirichlet and Thomson principles. The Dirichlet principle characterizes $\text{cap}(A)$ for a finite subset A of $\mathbb{Z}^d$ as the infimum over all Dirichlet forms of real-valued functions of finite support taking the value 1 on A, while the Thomson principle asserts that $1/\text{cap}(A)$ equals the infimum over all L_2-energies of unit flows on the edges of $\mathbb{Z}^d$ from the set A to infinity. According to Aldous and Fill [3], analogues on finite graphs of these variational principles characterize the expected time until the first visit of the random walk to a subset of vertices. The main part of the proof consists of the construction of optimizing objects for these variational problems.

2.4. Random walks on cylinders and random interlacements. In Chapter 4, we prove results similar to the one just described for random walks on discrete cylinders. The motivation for such results is the same as the one given in the previous section: if one can link the distribution of random walk trajectories on cylinders to random interlacements, then progress on percolation problems on random interlacements may improve the bounds on disconnection times of discrete cylinders. For disconnection times T_N of $(\mathbb{Z}/N\mathbb{Z})^d \times \mathbb{Z}$, $d \geq 2$, this idea has already been fruitfully applied by Sznitman in [36], [37] and [38]. Thanks to these works, it is known that the laws of T_N/N^{2d} on $(0, \infty)$ are tight.

We consider random walks on more general cylinders of the form $G_N \times \mathbb{Z}$ running up to a time of order $|G_N|^2$, where the bases G_N are given by large finite weighted graphs satisfying suitable assumptions. For many examples of graphs G_N, the timescale of order $|G_N|^2$

approximately corresponds to the asymptotic behavior of the disconnection time, as is known from [35]. The result we prove asserts that in the large N limit, the microscopic pictures of visited sites in distant neighborhoods are described by the model of random interlacements on transient weighted graphs of the form $\mathbb{G} \times \mathbb{Z}$. Sznitman's construction of random interlacements on $\mathbb{Z}^d$, $d \geq 3$, has been described in the beginning of the last section and can be extended to graphs of this form, see [40]. We thereby generalize a result proved by Sznitman, who considers the case $G_N = (\mathbb{Z}/N\mathbb{Z})^d$, $d \geq 2$, in [36]. Some additional examples of graphs G_N to which our result applies are for instance d-dimensional boxes with reflecting boundary of side-length N, discrete Sierpinski graphs of depth N, or d-ary trees of depth N for $d \geq 2$.

The random walk $(X_n)_{n \geq 0}$ on $G_N \times \mathbb{Z}$ is defined as a Markov chain on the weighted graph $G_N \times \mathbb{Z}$ equipped with the product graph structure. At each step, the random walk X_n moves from the current vertex to another vertex with a probability that is proportional to the weight of the edge between the two vertices, where we set the weight of every edge in the $\mathbb{Z}$-direction equal to $1/2$. We again consider M sequences of points

$$x_{m,N} = (y_{m,N}, z_{m,N}), \; 1 \leq m \leq M, \; \text{in } G_N \times \mathbb{Z},$$

with mutual distance tending to infinity as N tends to infinity. We also assume that balls centered at the points $y_{m,N} \in G_N$ of a radius diverging with N are isomorphic to balls in infinite graphs $\mathbb{G}_m$. This assumption allows the definition of the vacant configurations $\omega_t^m(.)$ in the neighborhoods of $x_{m,N}$ at time $t \geq 0$ in analogy with (2.4). An important difference to the result of Theorem 3 is that on the cylinder the dependence between local configurations in different neighborhoods does not vanish in the large N limit. To state the main result, we introduce the local time $(L_n^z)_{n \geq 1}$ of the $\mathbb{Z}$-projection of the random walk at site $z \in \mathbb{Z}$ and a jointly continuous version $\mathcal{L}(v,t)$, $v \in \mathbb{R}$, $t \geq 0$, of the local time of the canonical Brownian motion.

THEOREM 4. (*Theorem 1.1, Chapter 4, p. 105*) *Assume that the average weight per vertex in G_N converges to some number $\beta > 0$ and that*

$$\frac{z_{m,N}}{|G_N|} \longrightarrow v_m \in \mathbb{R}, \; \text{as } N \text{ tends to infinity, for } 1 \leq m \leq M.$$

Then under suitable additional assumptions on the graphs G_N (that hold for the examples mentioned above), the random variables

$$\left(\omega^1_{\alpha|G_N|^2}, \ldots, \omega^M_{\alpha|G_N|^2}, \frac{L^{z_{1,N}}_{\alpha|G_N|^2}}{|G_N|}, \ldots, \frac{L^{z_{M,N}}_{\alpha|G_N|^2}}{|G_N|} \right), \quad \alpha > 0, \; N \geq 1,$$

converge in joint distribution as N tends to infinity to the law of a random vector

$$(\omega_1, \ldots, \omega_M, U_1, \ldots, U_M).$$

This vector has the following distribution: $(U_m)_{m=1}^M$ is distributed as $((1+\beta)\mathcal{L}(v_m, \alpha/(1+\beta)))_{m=1}^M$ with β as above, and conditionally on $(U_m)_{m=1}^M$, the variables $(\omega_m)_{m=1}^M$ have joint distribution

$$\prod_{1 \leq m \leq M} \mathbb{Q}_{U_m/(1+\beta)}^{\mathbb{G}_m \times \mathbb{Z}},$$

where $\mathbb{Q}_u^{\mathbb{G}_m \times \mathbb{Z}}$ denotes the law on $\{0,1\}^{\mathbb{G}_m \times \mathbb{Z}}$ of the vacant set of the random interlacement at level $u \geq 0$ (as in (2.3) with $\mathbb{G}_m \times \mathbb{Z}$ in place of $\mathbb{Z}^d$).

In the proof of this result, we rely on estimates on the spectral gap of G_N and the heat kernel decay on G_N and $\mathbb{G}_m$, which allow us to compare the trace left by the random walk in the neighborhoods of $x_{m,N}$ to the trace of the independent paths of a random interlacement. One of the additional difficulties that arise with respect to the special case $G_N = (\mathbb{Z}/N\mathbb{Z})^d$ is that the $\mathbb{Z}$-projection of the random walk is in general not a Markov process. Indeed, the $\mathbb{Z}$-projection stays at every site visited for an amount of time that depends on the current position of the G_N-projection of the random walk. In order to overcome this difficulty, we decouple the $\mathbb{Z}$-component of the random walk from the G_N-component by introducing a continuous-time process $\mathsf{X} = (\mathsf{Y}, \mathsf{Z})$, such that the G_N- and $\mathbb{Z}$-components Y and Z are independent and such that the process X restricted to the jump times is the random walk X on $G_N \times \mathbb{Z}$. By assuming bounds on the spectral gap of G_N, we can prove an ergodic theorem for the process of jump times and thereby transfer results from continuous- to discrete time.

3. Organization of the thesis

This thesis consists of Chapters 1-4. These four chapters correspond to the four sections 2.1-2.4 of the Introduction and to the four articles [**43**], [**44**], [**45**] and [**46**], respectively.

CHAPTER 1

Disconnection of a discrete cylinder by a biased random walk

We consider a random walk on the discrete cylinder $(\mathbb{Z}/N\mathbb{Z})^d \times \mathbb{Z}$, $d \geq 3$ with drift $N^{-d\alpha}$ in the $\mathbb{Z}$-direction and investigate the large N-behavior of the disconnection time T_N, defined as the first time when the trajectory of the random walk disconnects the cylinder into two infinite components. We prove that, as long as the drift exponent α is strictly greater than 1, the asymptotic behavior of T_N remains $N^{2d+o(1)}$, as in the unbiased case considered by Dembo and Sznitman, whereas for $\alpha < 1$, the asymptotic behavior of T_N becomes exponential in N.

1. Introduction

Informally, the object of our study can be described as follows: a particle feeling a drift moves randomly through a cylindrical object, and damages every visited point. How long does it take until the cylinder breaks apart, and how does the answer to this question depend on the drift felt by the particle? This is a variation on the problem of "the termite in a wooden beam" considered by Dembo and Sznitman [12].

We henceforth consider the *discrete cylinder*

(1.1) $$E = \mathbb{T}_N^d \times \mathbb{Z}, \quad d \geq 1,$$

where $\mathbb{T}_N^d$ denotes the d-dimensional integer torus $\mathbb{T}_N^d = (\mathbb{Z}/N\mathbb{Z})^d$. The disconnection time of the cylinder E by a simple (unbiased) random walk was introduced by Dembo and Sznitman in [12], where it was shown that its asymptotic behavior is approximately $N^{2d} = |\mathbb{T}_N^d|^2$ as $N \to \infty$ when $d \geq 1$. This result was extended by Sznitman in [35] to a wide class of bases of E with uniformly bounded degree as $N \to \infty$. Similar models related to interfaces created by simple random walk trajectories have been studied by Benjamini and Sznitman [6] and Sznitman [36]. The former of these two works has led Dembo and Sznitman [13] to sharpen their lower bound on the disconnection time of E for large d. Here we investigate the disconnection time for a random walk with bias into the $\mathbb{Z}$-direction.

We now proceed to the precise description of the problem studied in the present work. The cylinder E is equipped with the Euclidean distance $|.|$ and the natural product graph structure, for which all vertices $x_1, x_2 \in E$ with $|x_1 - x_2| = 1$ are connected by an edge. The (discrete-time) random walk with drift $\Delta \in [0,1)$ is the Markov chain $(X_n)_{n \geq 0}$ on E with starting point $x \in E$ and transition probability

(1.2) $\quad p_X(x_1, x_2) = \dfrac{1 + \Delta(\pi_{\mathbb{Z}}(x_2 - x_1))}{2d + 2} \mathbf{1}_{\{|x_1 - x_2| = 1\}}, \quad x_1, x_2 \in E,$

where $\pi_{\mathbb{Z}}$ denotes the projection from E onto $\mathbb{Z}$. The process is defined on a suitable filtered probability space $(\Omega_N, (\mathcal{F}_n)_{n \geq 0}, P_x^{\Delta})$ (see Section 2 for details). In particular, under P_0^0, X is the ordinary simple random walk on E. We say that a set $K \subseteq E$ *disconnects* E if $\mathbb{T}_N^d \times (-\infty, -M]$ and $\mathbb{T}_N^d \times [M, \infty)$ are contained in two distinct components of $E \setminus K$ for large $M \geq 1$. The central object of interest is the *disconnection time*

(1.3) $\quad T_N = \inf\{n \geq 0 : X([0,n]) \text{ disconnects } E\}.$

We consider drifts of the form $N^{-d\alpha} = |\mathbb{T}_N^d|^{-\alpha}$, $\alpha > 0$. Our main result shows that the asymptotic behavior of T_N as $N \to \infty$ is the same as in the case without drift considered in [**12**] as long as $\alpha > 1$, and becomes exponential in N when $\alpha < 1$:

THEOREM 1.1. $(d \geq 3, \alpha > 0, \epsilon > 0)$

(1.4) $\quad \begin{array}{ll} \text{If } \alpha > 1, & N^{2d-\epsilon} \leq T_N \leq N^{2d+\epsilon}, \\ \text{if } \alpha < 1, & \exp\{N^{d(1-\alpha-\varphi(\alpha))-\epsilon}\} \leq T_N \leq \exp\{N^{d(1-\alpha)+\epsilon}\}, \end{array}$

with probability tending to 1 *as* $N \to \infty$, *where the piecewise linear function* $\varphi : (0,1) \to \left(0, \frac{1}{d-1}\right)$ *is defined by*

(1.5) $\quad \varphi(\alpha) = \alpha \mathbf{1}_{\{0 < \alpha < \alpha_*\}} + \left(\dfrac{1}{d} + \dfrac{\alpha}{d-1} - \alpha\right) \mathbf{1}_{\{\alpha_* < \alpha < \frac{1}{d}\}}$

$\qquad\qquad + \dfrac{1-\alpha}{(d-1)^2} \mathbf{1}_{\{\frac{1}{d} \leq \alpha < 1\}},$

for $\alpha_* = \frac{1}{d(2-\frac{1}{d-1})}$. *In particular,* φ *satisfies* $\lim_{\alpha \to 0} \varphi(\alpha) = \lim_{\alpha \to 1} \varphi(\alpha) = 0$ *(see Figure 1 for an illustration of the region between* $1 - \alpha - \varphi(\alpha)$ *and* $1 - \alpha$*).*

We now outline the ideas entering the proof of this result. The upper bounds on T_N are derived in Theorem 3.1. The proof of this theorem is based on the simple observation that the cylinder E is disconnected as soon as a slice of the form $\mathbb{T}_N^d \times \{z\} \subseteq E$ is completely covered by the walk. We thus show that the trajectory of the random walk X up to time $N^{2d+\epsilon}$ (for $\alpha > 1$), or $\exp\{N^{d(1-\alpha)+\epsilon}\}$ (for $\alpha < 1$), does cover such a slice with probability tending to 1 as $N \to \infty$. To this end, we fix the slice $\mathbb{T}_N^d \times \{0\}$ and record visits made to it by X. The process recording these visits is defined as $(V, 0)$ (cf. (3.4)). Once we have checked that V forms a Markov chain on $\mathbb{T}_N^d$ in Lemma 3.3, we can infer from the coupon-collector-type estimate (3.5) on the cover time that after a certain "critical" number of visits, the slice $\mathbb{T}_N^d \times \{0\}$ is covered with overwhelming probability by $(V, 0)$, hence by X. Since

1. Introduction

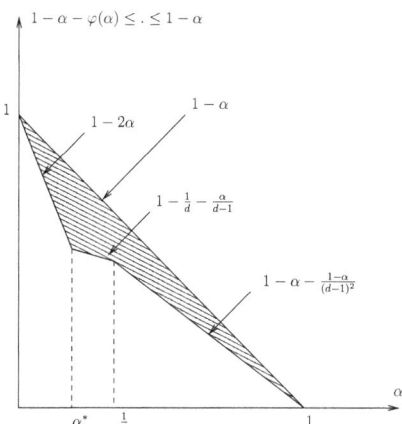

Figure 1: The shaded region lies between the exponents of the upper and lower bounds in Theorem 1.1 for $\alpha \in (0,1)$.

the same estimates apply to any slice $\mathbb{T}_N^d \times \{z\}$, $z \in \mathbb{Z}$, we are left with the one-dimensional problem of finding an upper bound on the time until sufficiently many such visits occur for some slice $\mathbb{T}_N^d \times \{z\}$.

Let us now describe the ideas involved in the more delicate derivation of the lower bounds. In this work, we reduce the problem of finding a lower bound on T_N to a large deviation problem concerning the disconnection of a certain finite subset of E by excursions of an unbiased simple random walk, and then derive estimates on this large deviation problem. Let us describe this last problem and the reduction step in more detail. For any subsets K, $B \subseteq E$, B finite, and $\kappa \in \left(0, \frac{1}{2}\right)$, we say that K *κ-disconnects* B if K contains the relative boundary in B of a subset of B with relative volume between κ and $1 - \kappa$, i.e. if there is a subset I of B (generally not unique) such that

(1.6) $\qquad \kappa |B| \leq |I| \leq (1-\kappa)|B|$ and $\partial_B(I) \subseteq K$,

where, for sets $A, B \subseteq E$, $|A|$ denotes the number of points in A and $\partial_B(A)$ the B-relative boundary of A, i.e. the set of points in $B \setminus A$ with neighbors in A. The set whose disconnection concerns us is

(1.7) $\qquad B(\alpha) = \left[-\left[\frac{N}{4}\right], \left[\frac{N}{4}\right] \right]^d \times \left[-\left[\frac{N^{d\alpha \wedge 1}}{4}\right], \left[\frac{N^{d\alpha \wedge 1}}{4}\right] \right].$

Note that in the case $\alpha \geq \frac{1}{d}$, $B(\alpha)$ becomes $B_\infty(0, [N/4])$, the closed ball of radius $[N/4]$ with respect to the l_∞-distance, centered at 0. We define $U_{B(\alpha)}$ as the first time when the trajectory of the random walk

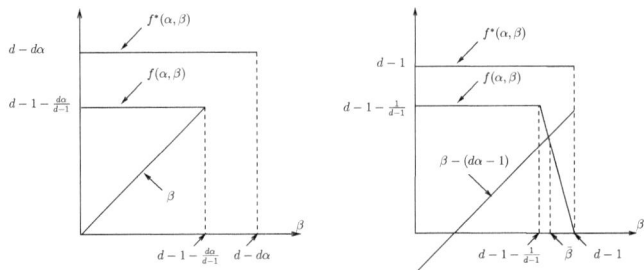

Figure 2: The function f provided by Theorem 6.1, case $\alpha \in (0, \frac{1}{d})$ on the left, case $\alpha \in [\frac{1}{d}, \infty)$ on the right.

$\frac{1}{3}$-disconnects $B(\alpha)$, that is

(1.8) $\quad U_{B(\alpha)} = \inf \left\{ n \geq 0 : X([0,n]) \; \frac{1}{3}\text{-disconnects } B(\alpha) \right\}.$

The random walk excursions featuring in the large deviation problem are excursions in and out of slices of the form

(1.9) $\quad S_r = \mathbb{T}_N^d \times \bigl[-[r],[r]\bigr] \subseteq E \quad (r > 0).$

Finally, the crucial reduction step comes in the following theorem, proved in Section 4:

THEOREM 1.2. $(d \geq 2, \; \alpha > 0, \; \beta > 0)$ Suppose that f is a non-negative function on $(0, \infty)^2$ such that, for $(\mathcal{R}_n)_{n \geq 1}$, $(\mathcal{D}_n)_{n \geq 1}$, the successive returns to $S_{2[N^{d\alpha} \wedge 1]}$ and departures from $S_{4[N^{d\alpha} \wedge 1]}$ (cf. (2.19)) and the stopping time defined in (1.8) one has

(1.10) $\quad \varlimsup_{N \to \infty} \frac{1}{N^\xi} \log \sup_{x \in S_{2[N^{d\alpha} \wedge 1]}} P_x^0 \bigl[U_{B(\alpha)} \leq \mathcal{D}_{[N^\beta]} \bigr] < 0,$

for $0 < \xi < f(\alpha, \beta)$.

If $f(\alpha, \beta) > 0$ for all $\alpha > 1$, $\beta \in (0, d-1)$, then it follows that

(1.11) $\quad P_0^{N^{-d\alpha}}[N^{2d-\epsilon} \leq T_N] \xrightarrow{N \to \infty} 1, \quad \text{for any } \alpha > 1, \; \epsilon > 0,$

while for any $f \geq 0$,

(1.12) $\quad P_0^{N^{-d\alpha}}\bigl[\exp\{N^{\zeta - \epsilon}\} \leq T_N\bigr] \xrightarrow{N \to \infty} 1, \quad \text{for any } \alpha > 0, \; \epsilon > 0,$

where

(1.13) $\quad \zeta = \sup_{\beta > 0} g_\alpha(\beta) \quad \text{and} \quad g_\alpha(\beta) = \bigl(\beta - (d\alpha - 1)_+\bigr) \wedge f(\alpha, \beta).$

1. Introduction

In order to apply Theorem 1.2, one has to find a suitable nonnegative function f satisfying the fundamental large deviation estimate (1.10). We show in Theorem 6.1 that (1.10) holds for the function f illustrated in Figure 2. With this function f, the lower bound exponents ζ (in (1.12)) and $d(1 - \alpha - \varphi(\alpha))$ (in (1.4)) are related via

$$(1.14) \qquad d(1 - \alpha - \varphi(\alpha)) = \zeta \vee \left(d(1 - 2\alpha)\mathbf{1}_{\{\alpha < \frac{1}{d}\}} \right),$$

as is shown in Corollary 6.3. The fact that the lower bound on T_N holds with the expression $d(1 - 2\alpha)\mathbf{1}_{\{\alpha < \frac{1}{d}\}}$ in (1.14) follows from the rather straightforward lower bound derived in Proposition 6.2.

We now sketch some of the techniques involved in the proof of Theorem 1.2 and the subsequent derivation of the large deviation estimate (1.10). The first step in the proof of Theorem 1.2 is a purely geometric argument in the spirit of Dembo and Sznitman [12] showing that any trajectory disconnecting E must $\frac{1}{3}$-disconnect a set of the form $x_* + B(\alpha)$ (see Lemma 4.1). On the event that the walk performs no more than $[N^\beta]$ excursions between $x_* + S_{2[N^{d\alpha-1}]}$ and $x_* + S^c_{4[N^{d\alpha-1}]}$ for any $x_* \in E$ until some time t_N, disconnection before time t_N can only occur if these at most $[N^\beta]$ excursions $\frac{1}{3}$-disconnect $x_* + B(\alpha)$ for some $x_* \in E$. One can thus apply the assumed large deviation estimate (1.10) after getting rid of the drift with the help of a Girsanov-type control (see Lemma 2.1) and applying translation invariance. It then remains to bound the probability that more than $[N^\beta]$ of the above-mentioned excursions occur for some $x_* \in E$. This can be achieved with standard estimates on one-dimensional random walk.

In order to derive the fundamental large deviation estimate (1.10), we begin with some more geometric lemmas. We show in Lemmas 5.1-5.3 that when $0 < \gamma < \gamma' < 1$, for large N and any set K $\frac{1}{3}$-disconnecting $B_\infty(0, [N/4])$ (cf. (1.7) and thereafter), one can find a subcube of $B_\infty(0, [N/4])$ with size $L = [N^{\gamma'}]$, so that K contains a large set of points in each of a "well-spread" collection of sub-subcubes with size $l = [N^\gamma]$ (we refer to Lemma 5.3 for the precise statement). A key ingredient for the proof of this geometric result, similar to Lemma 2.5 of [12], is an isoperimetric inequality from [14], see Lemma 5.2. A small modification of the argument shows a similar result for $B(\alpha)$, $\alpha < \frac{1}{d}$ (see Lemma 5.4). As a consequence of these geometric results, one finds that the event under consideration in the large deviation estimate (1.10) is included in the event that the trajectory left by the $[N^\beta]$ excursions has substantial presence in many small subcubes of $B(\alpha)$. The key control on an event of this form is provided by Lemma 6.5. The main part of the argument there is to obtain a tail estimate on the number of points contained in the projection on one of the d-dimensional hyperplanes of the small subcubes intersected with the trajectory of the random walk stopped when exiting a large set. It follows from Khaśminskii's Lemma that this number of points, divided by its expectation, is a random variable whose exponential moment is uniformly bounded with N. In order to bound the expected number of visited points, we use standard estimates on the Green function of the simple random walk.

An obvious question arising from Theorem 1.1 is whether one can prove the same result with $\varphi \equiv 0$ in (1.4). With Theorem 1.2, it is readily seen that this would follow if one could show that the large deviation estimate (1.10) holds with

$$(1.15) \qquad f^*(\alpha, \beta) = \begin{cases} d - (d\alpha \wedge 1) & \beta < d - (d\alpha \wedge 1), \\ 0 & \beta \geq d - (d\alpha \wedge 1), \end{cases}$$

see Figure 2. In fact, the above function f^* can be shown to be the correct exponent associated to a large deviation problem similar to (1.10), where one replaces the time $U_{B(\alpha)}$ by $\mathcal{U}$, defined as the first time when the trajectory of X covers $\mathbb{T}_N^d \times \{0\}$. Plainly one has $U_{B(\alpha)} \leq \mathcal{U}$, and it follows that any function f in (1.10) satisfies $f(\alpha, \beta) \leq f^*(\alpha, \beta)$ for all points (α, β) of continuity of f, we refer to Remark 6.7 for more details. The crucial open question is therefore: are these two problems sufficiently similar for (1.10) to hold with f^*?

Organization of the article. In Section 2, we provide the definitions and the notation to be used throughout this article and prove a Girsanov-type estimate to be frequently used later on.

In Section 3, we derive the upper bounds on T_N of Theorem 1.1.

In Section 4, we prove Theorem 1.2, thus reducing the derivation of a lower bound on T_N to a large deviation estimate.

In Section 5, we prove several geometric lemmas in preparation of our derivation of the latter estimate.

In Section 6, we supply the key large deviation estimate in Theorem 6.1 and derive a simple lower bound on T_N for large drifts. As we show, this yields the lower bounds on T_N in Theorem 1.1.

Constants. Finally, we use the following convention concerning constants: Throughout the text, c or c' denote strictly positive constants which only depend on the base-dimension d, with values changing from place to place. The numbered constants $c_0, c_1, \ldots$ are fixed and refer to their first place of appearance in the text. Dependence of constants on parameters other than d appears in the notation. For example, $c(\gamma, \gamma')$ denotes a positive constant depending on d, γ and γ'.

Acknowledgments. The author is indebted to Alain-Sol Sznitman for suggesting the problem and for fruitful advice throughout the completion of this work. Thanks are also due to Laurent Goergen for pertinent remarks on a previous version of this article.

2. Definitions, notation and a useful estimate

The purpose of this section is to set up the notation and the definitions to be used in this article and to provide a Girsanov-type estimate comparing the random walks with drift and without drift, to be frequently applied later on.

Throughout this article, we denote, for $s, t \in \mathbb{R}$, by $s \wedge t$ the minimum of s and t, by $s \vee t$ the maximum of s and t, by $[s]$ the largest integer satisfying $[s] \leq s$ and we set $t_+ = t \vee 0$ and $t_- = -(t \wedge 0)$. Recall that we introduced the cylinder E in (1.1). E is equipped with the Euclidean distance $|.|$ and the l_∞-distance $|.|_\infty$. We denote a generic element of E by $x = (u, v)$, $u \in \mathbb{T}_N^d$, $v \in \mathbb{Z}$ and the corresponding closed ball of $|.|_\infty$-radius $r > 0$ centered at $x \in E$ by $B_\infty(x, r)$. Note that E is the image of $\mathbb{Z}^{d+1} = \mathbb{Z}^d \times \mathbb{Z}$ by the mapping $\pi_E \colon \mathbb{Z}^d \times \mathbb{Z} \to E$, $(u, v) \mapsto (\pi_N(u), v)$, where π_N denotes the canonical projection from $\mathbb{Z}^d$ onto the torus $\mathbb{T}_N^d$. We write $\{e_i\}_{i=1}^{d+1}$ for the canonical basis of $\mathbb{R}^{d+1}$. The projections π_i, $i = 1, \ldots, d+1$ onto the d-dimensional hyperplanes of E are the mappings from E to $(\mathbb{Z}/N\mathbb{Z})^{d-1} \times \mathbb{Z}$ when $i = 1, \ldots, d$, or to $(\mathbb{Z}/N\mathbb{Z})^d$ when $i = d+1$, defined by omitting the i-th component of $(u, v) = (u^1, \ldots, u^d, v) \in E$. These projections are not to be confused with the projection $\pi_\mathbb{Z}$ from E onto $\mathbb{Z}$, defined by

$$\pi_\mathbb{Z}((u, v)) = v. \tag{2.1}$$

For any subset $A \subseteq E$ and $l \geq 1$, we define the l-neighborhood of A,

$$A^{(l)} = \{x \in E : \text{ for some } x' \in A, |x - x'|_\infty \leq l\}, \tag{2.2}$$

its l-interior,

$$A^{(-l)} = \{x \in A : \text{ for all } x' \notin A, |x - x'|_\infty > l\} \tag{2.3}$$

(so that $A \subseteq B^{(-l)}$ if and only if $A^{(l)} \subseteq B$) and its diameter

$$\operatorname{diam}(A) = \sup\{|x - x'|_\infty : x, x' \in A\}. \tag{2.4}$$

Given another subset $B \subseteq E$, we define the B-relative boundary of A,

$$\partial_B(A) = \{x \in B \setminus A : \text{ for some } x' \in A, |x - x'| = 1\}, \tag{2.5}$$

and the B-relative boundary of A in direction $i \in \{1, \ldots, d+1\}$,

$$\partial_{B,i}(A) = \{x \in B \setminus A : \text{ for some } x' \in A, |x - x'| = 1 \tag{2.6}$$
$$\text{ and } \pi_i(x) = \pi_i(x')\}.$$

The cube of side-length $l - 1$, $l = 1, \ldots, N$ is defined as

$$C(l) = [0, l-1]^{d+1} \subseteq E \tag{2.7}$$

(where $[0, l-1] = \{0, \ldots l-1\}$) and the same cube with base-point $x \in E$ as

$$C_x(l) = x + C(l), \tag{2.8}$$

where, for $x \in E$ and $A \subset E$, we set $x + A = \{x + x' : x' \in A\} \subseteq E$.

We define the process $(X_n)_{n\geq 0}$ as the canonical coordinate process on the space of nearest-neighbor paths on E with infinitely many jumps in both the $\mathbb{T}_N^d$ and the $\mathbb{Z}$ directions. The process X generates the canonical filtration $(\mathcal{F}_n)_{n\geq 0}$ and has the associated shift operators $(\theta_n)_{n\geq 0}$. We define the process of jump times $(J_k^Y)_{k\geq 0}$ corresponding to the projection of X on $\mathbb{T}_N^d$ by

$$J_0^Y = 0, J_1^Y = \inf\{n \geq 1 : \pi_{d+1}(X_n) \neq \pi_{d+1}(X_{n-1})\} \in \{1, 2, \ldots\}$$
$$J_k^Y = J_1^Y \circ \theta_{J_{k-1}^Y} + \theta_{J_{k-1}^Y}, \text{ for } k \geq 2.$$

The process $(J_k^Z)_{k\geq 0}$ is defined analogously, with π_{d+1} replaced by $\pi_\mathbb{Z}$. The process $(\rho_k^Y)_{k\geq 0}$ is defined as the process counting the number of jumps of $\pi_{d+1}(X)$ until time n, i.e.

(2.9) $\rho_n^Y = \sup\{k \geq 1 : J_k^Y \leq n\}$, for $n \geq 0$, where we set $\sup \emptyset = 0$.

Analogously, we define $(\rho_k^Z)_{k\geq 0}$ with J^Y replaced by J^Z. We can then define the processes $(Y_n)_{n\geq 0}$ and $(Z_n)_{n\geq 0}$ on $\mathbb{T}_N^d$ and $\mathbb{Z}$ by

(2.10) $\quad Y_n = \pi_{d+1}(X_{J_n^Y})$, and $Z_n = \pi_\mathbb{Z}(X_{J_n^Z})$, $n \geq 0$,

so that we have

(2.11) $\quad X_n = (Y_{\rho_n^Y}, Z_{\rho_n^Z})$, for $n \geq 0$,

as well as

(2.12) $\quad J_k^Y = \inf\{n \geq 0 : \rho_n^Y = k\}$, for $k \geq 0$,

and analogously for J_k^Z.

We then construct the probability measures P_x^Δ, for $x = (u,v) \in E$ and $0 \leq \Delta < 1$ on $(E^\mathbb{N}, (\mathcal{F}_n)_{n\geq 0})$ (and write E_x^Δ for the corresponding expectations) such that, under P_x^Δ,

(2.13)

the processes $(Y_n)_{n\geq 0}$, $(Z_n)_{n\geq 0}$, and $(\rho_n^Y, \rho_n^Z)_{n\geq 0}$ are independent,

(2.14) $\quad Y$ is a simple random walk on $\mathbb{T}_N^d$ with starting point u,

(2.15) $\quad Z$ is a random walk on $\mathbb{Z}$ starting at v with transition probability $p_Z(v', v'-1) = \frac{1-\Delta}{2}$, $p_Z(v', v'+1) = \frac{1+\Delta}{2}$,

for $v' \in \mathbb{Z}$ (so Δ can be interpreted as the drift of the walk in the $\mathbb{Z}$-component),

(2.16) $\quad \left(\rho_n^Y - \rho_{n-1}^Y, \rho_n^Z - \rho_{n-1}^Z\right)_{n\geq 0}$ is a sequence of iid random variables with distribution $\frac{d}{d+1}\delta_{(1,0)} + \frac{1}{d+1}\delta_{(0,1)}$.

It follows from this construction that, under P_x^Δ, X is a random walk on E with drift Δ starting at x, i.e. a Markov chain on E with initial distribution $\delta_{\{x\}}$ and transition probability specified in (1.2) (in

2. Definitions, notation and a useful estimate

particular, the notation P_x^Δ, $x \in E$, is consistent with its use in the introduction). We will frequently use the following stopping times: The entrance and hitting times H_A^X and $\tilde{H}_A^X$ of the set $A \subseteq E$,

(2.17) $\quad H_A^X = \inf\{n \geq 0; X_n \in A\}$, and $\tilde{H}_A^X = \inf\{n \geq 1; X_n \in A\}$

where we write H_x^X if $A = \{x\}$, and the cover time C_A^X of $A \subseteq E$,

(2.18) $\quad\quad\quad C_A^X = \inf\{n \geq 0; X([0,n]) \supseteq A\},$

with obvious modifications such as H^Z for processes other than X. For the random walk X and any sets $A \subseteq \bar{A} \subseteq E$, the successive returns $(R_n)_{n\geq 1}$ to A and departures $(D_n)_{n\geq 1}$ from $\bar{A}$ are defined as

(2.19) $\quad R_1 = H_A^X, \quad D_1 = H_{\bar{A}^c}^X \circ \theta_{R_1} + R_1$ and for $n \geq 2$,
$\quad\quad\quad R_n = R_1 \circ \theta_{D_{n-1}} + D_{n-1}, \quad D_n = D_1 \circ \theta_{D_{n-1}} + D_{n-1},$

so that $0 \leq R_1 \leq D_1 \leq \cdots \leq R_n \leq D_n \leq \cdots \leq \infty$ and P_x^Δ-a.s. all these inequalities are strict, except possibly the first one. Finally, we also use the Green function of the simple random walk X without drift, killed when exiting $A \subseteq E$, defined as

(2.20) $\quad g^A(x, x') = E_x^0\left[\sum_{n=0}^{\infty} \mathbf{1}\{X_n = x', n < H_{A^c}^X\}\right], \quad x, x' \in E.$

We conclude this section with the Girsanov-type estimate comparing P_x^Δ and P_x^0.

LEMMA 2.1. ($d \geq 1$, $N \geq 1$, $\Delta \in (0,1)$, $x \in E$)
Consider any $(\mathcal{F}_n)_{n\geq 0}$-stopping time T and any $\mathcal{F}_T$-measurable event A such that, for some $b, b' \in \mathbb{R} \cup \{-\infty, \infty\}$,

(2.21) $\quad T < \infty \quad$ and $\quad b \leq \pi_\mathbb{Z}(X_T - x) \leq b' \quad P_x^0$-a.s. on A. Then

(2.22) $\quad (1-\Delta)^{b_-}(1+\Delta)^{b_+} E_x^0[A, (1-\Delta^2)^{\lfloor \frac{T}{2} \rfloor}] \leq P_x^\Delta[A] \quad$ and

(2.23) $\quad\quad\quad P_x^\Delta[A] \leq (1-\Delta)^{b'_-}(1+\Delta)^{b'_+} P_x^0[A],$

where we set $(1-\Delta)^\infty = 0$ and $(1+\Delta)^\infty = \infty$.

PROOF. For any $\mathcal{F}_n$-measurable event A_n, it follows directly from the definition of the transition probabilities of the walk X (cf. (1.2)) that

(2.24) $\quad\quad P_x^\Delta[A_n] = E_x^0\left[A_n, \prod_{i=1}^n (1 + \Delta \pi_\mathbb{Z}(X_i - X_{i-1}))\right].$

For any $(\mathcal{F}_n)_{n\geq 0}$-stopping time T satisfying (2.21), we apply (2.24) with the $\mathcal{F}_n$-measurable event $A_n = A \cap \{T = n\}$ for $n \geq 0$ and deduce, via monotone convergence,

$$(2.25) \quad P_x^\Delta[A] = \sum_{n \geq 0} P_x^\Delta[A_n]$$

$$= \sum_{n \geq 0} E_x^0 \left[A_n, \prod_{i=1}^T (1 + \Delta \pi_\mathbb{Z}(X_i - X_{i-1})) \right]$$

$$= E_x^0 \left[A, \prod_{i=1}^T (1 + \Delta \pi_\mathbb{Z}(X_i - X_{i-1})) \right].$$

To complete the proof, we bound the product inside the expectation on the right-hand side of (2.25) from above and from below. The contribution of the product is a factor of $1 + \Delta$ for every displacement of X into the positive $\mathbb{Z}$-direction up to time T and a factor of $1 - \Delta$ for every displacement into the negative $\mathbb{Z}$-direction during the same time. We now group together the factors in the product as pairs of the form $(1 + \Delta)(1 - \Delta) = 1 - \Delta^2$ for as many factors as possible (i.e. until all remaining factors are of the form $1 + \Delta$ or all remaining factors are of the form $1 - \Delta$). By (2.21), the contribution of these remaining factors is bounded from below by $(1 - \Delta)^{b_-}(1 + \Delta)^{b_+}$ and from above by $(1 - \Delta)^{b'_-}(1 + \Delta)^{b'_+}$. For (2.23), we note that $1 - \Delta^2 < 1$ and bound the contribution made by the pairs from above by 1. For (2.22), we note that the number of pairs contributed can be at most $\left[\frac{T}{2}\right]$. This completes the proof of the lemma. □

3. Upper bound

This section is devoted to upper bounds on T_N. We will prove the following theorem, which is more than sufficient to yield the upper bounds in Theorem 1.1:

THEOREM 3.1. ($d \geq 2$, $\alpha > 0$, $\epsilon > 0$)
For some constant $c_0 > 0$,

$$(3.1) \quad \text{for } \alpha > 1, \quad P_0^{N^{-d\alpha}}\left[T_N \leq N^{2d}(\log N)^{4+\epsilon}\right] \overset{N \to \infty}{\longrightarrow} 1,$$

$$(3.2) \quad \text{for } \alpha \leq 1, \quad P_0^{N^{-d\alpha}}\left[T_N \leq \exp\{c_0 N^{d(1-\alpha)}(\log N)^2\}\right] \overset{N \to \infty}{\longrightarrow} 1.$$

PROOF. Following the idea outlined in the introduction, we define the process V, whose purpose is to record visits of X to $\mathbb{T}_N^d \times \{0\}$. To this end, we introduce the stopping times $(S_n)_{n \geq 0}$ by setting (cf. (2.17))

$$(3.3) \quad S_0 = 0, \text{ and for } n \geq 1,$$

$$S_n = \begin{cases} \tilde{H}_{\mathbb{T} \times \{0\}}^X \circ \theta_{S_{n-1}} + S_{n-1} & \text{on } \{S_{n-1} < \infty\}, \\ \infty & \text{on } \{S_{n-1} = \infty\}, \end{cases}$$

and on the event $\{S_k < \infty\}$, we define

$$(3.4) \quad V_n = Y_{\rho_{S_n}^Y} \overset{(2.11)}{=} \pi_{d+1}(X_{S_n}), \quad n = 0, \ldots, k.$$

3. Upper bound

Note that, as soon as V has visited all points of $\mathbb{T}_N^d$, X has visited all points of $\mathbb{T}_N^d \times \{0\}$, and has therefore disconnected E. Hence, we are interested in an upper bound on the cover time $C_{\mathbb{T}_N^d}^V$ (cf. (2.18)). This desired upper bound will result from the following estimate on cover times for symmetric Markov chains. Following Aldous and Fill [3], Chapter 7 (p. 2), we call a Markov chain $(W_n)_{n \geq 0}$ on the finite state-space G with transition probabilities $p_W(g, g')$, $g, g' \in G$ *symmetric*, if for any states $g_0, g_1 \in G$, there exists a bijection $\gamma : G \to G$ satisfying $\gamma(g_0) = g_1$ and $p_W(g, g') = p_W(\gamma(g), \gamma(g'))$ for all $g, g' \in G$.

LEMMA 3.2. *Given a symmetric, irreducible and reversible Markov chain $(W_n)_{n \geq 0}$ on the finite state-space G whose transition matrix*

$$(p_W(g, g'))_{g,g' \in G}$$

has eigenvalues $1 = \lambda_1(W) > \lambda_2(W) \geq \ldots \geq \lambda_{|G|}(W) \geq -1$, one has

$$(3.5) \qquad P_g\left[C_G^W \geq n\right] \leq |G| \exp\left\{-\left[\frac{n}{4eu(W)}\right]\right\},$$

for any $g \in G$, $n \geq 1$, where P_g is the canonical probability on $G^{\mathbb{N}}$ governing W with $W_0 = g$ and

$$(3.6) \qquad u(W) = \sum_{m=2}^{|G|} \frac{1}{1 - \lambda_m(W)}.$$

PROOF. We assume that $n \geq 4eu(W)$, for otherwise there is nothing to prove. The following estimate on the maximum hitting time (cf. (2.17)) is a consequence of the so-called eigentime identity (see [3], Lemma 15 and Proposition 13 in Chapter 3, and note that $E_g[H_{g'}^W] = E_{g'}[H_g^W]$ by our assumptions on symmetry, irreducibility and reversibility, cf. [3], Chapter 3, Lemma 1):

$$(3.7) \qquad \max_{g,g' \in G} E_g\left[H_{g'}^W\right] \leq 2 \sum_{m=2}^{|G|} \frac{1}{1 - \lambda_m(W)} \stackrel{(3.6)}{=} 2u(W).$$

Choosing any $1 \leq s \leq n$, we deduce the following tail estimate on C_G^W with a standard application of the simple Markov property at the times $([s] - 1)\left[\frac{n}{s}\right], \ldots, 2\left[\frac{n}{s}\right], \left[\frac{n}{s}\right]$:

$$P_g\left[C_G^W \geq n\right] = P_g\left[\text{for some } g' \in G : H_{g'}^W \geq n\right]$$

$$\leq |G| \max_{g,g' \in G} P_g\left[H_{g'}^W \geq n\right] \stackrel{(\text{Markov})}{\leq} |G| \left(\max_{g,g' \in G} P_g\left[H_{g'}^W \geq \left[\frac{n}{s}\right]\right]\right)^{[s]}$$

$$\stackrel{(\text{Chebychev}, (3.7))}{\leq} |G| \left(\left[\frac{n}{s}\right]^{-1} 2u(W)\right)^{[s]} \stackrel{(\frac{n}{s} \leq 2[\frac{n}{s}])}{\leq} |G| \left(\frac{4su(W)}{n}\right)^{[s]}.$$

With $1 \leq s = \frac{n}{4eu(W)} \leq n$, this yields (3.5). □

In order to apply Lemma 3.2, we now show that $(V_n)_{n=1}^k$ (cf. (3.4)) satisfies the hypotheses imposed on W, provided we take the event $\{S_k < \infty\}$ as probability space, equipped with the probability measure $P_{(u,0)}^\Delta [\cdot | S_k < \infty]$, $u \in \mathbb{T}_N^d$.

LEMMA 3.3. *($d \geq 1$, $k \geq 1$, $\Delta \in [0,1)$, $u \in \mathbb{T}_N^d$) On the probability space $\left(\{S_k < \infty\}, P_{(u,0)}^\Delta [\cdot | S_k < \infty]\right)$ and the finite time interval $n = 0, \ldots, k$, $(V_n)_{n=0}^k$ is a symmetric, irreducible and reversible Markov chain on $\mathbb{T}_N^d$ starting at u with transition probability*

$$(3.8) \qquad p_V(u, u') = P_{(u,0)}^\Delta \left[Y_{\rho_{S_1}^Y} = u' \big| S_1 < \infty \right], \quad u, u' \in \mathbb{T}_N^d.$$

PROOF. By (2.13) and the fact that S_1 and $\rho_{S_1}^Y$ are both $\sigma(Z, \rho^Y, \rho^Z)$-measurable, it follows that Y and $(\rho_{S_1}^Y, S_1)$ are independent. Hence, one can rewrite the expression for $p_V(u, u')$ in (3.8) using Fubini's theorem:

$$(3.9) \qquad p_V(u, u') = \frac{1}{P_0^\Delta [S_1 < \infty]} P_{(u,0)}^\Delta \left[Y_{\rho_{S_1}^Y} = u', S_1 < \infty \right]$$

$$\stackrel{\text{(Fubini)}}{=} \frac{1}{P_0^\Delta [S_1 < \infty]} E_{(u,0)}^\Delta \left[P_{(u,0)}^\Delta [Y_n = u'] \big|_{n = \rho_{S_1}^Y}, S_1 < \infty \right]$$

$$= \frac{1}{P_0^\Delta [S_1 < \infty]} E_0^\Delta \left[P_{(u,0)}^\Delta [Y_n = u'] \big|_{n = \rho_{S_1}^Y}, S_1 < \infty \right],$$

where in the last line we have used that the expression inside the expectation is a function of $\rho_{S_1}^Y$ and S_1 and therefore does not depend on the $\mathbb{T}_N^d$-coordinate of the starting point. From (3.9), it follows that the transition probabilities $p_V(\cdot, \cdot)$ define an irreducible, symmetric (as defined above Lemma 3.2) and reversible process. Indeed, for any $u, u' \in \mathbb{T}_N^d$ such that $P_{(u,0)}^\Delta [Y_1 = u'] > 0$, (3.9) and $P_0^\Delta [\rho_{S_1}^Y = 1, S_1 < \infty] \geq P_0^\Delta [X_1 \in \mathbb{T}_N^d \times \{0\}] > 0$ imply that $p_V(u, u') > 0$, so that irreducibility follows from irreducibility of the simple random walk Y. Similarly, (3.9) shows that symmetry follows from symmetry of Y, which holds by translation invariance. Finally, reversibility follows by exchanging u and u' in the last line of (3.9), which one can do by reversibility of Y. It thus remains to be shown that $p_V(\cdot, \cdot)$ are in fact the correct transition probabilities for V, i.e. that for any $u, u_1, \ldots, u_n \in \mathbb{T}_N^d$, $1 \leq n \leq k$, and

$$(3.10) \qquad A = \{V_0 = u, \ldots, V_{n-1} = u_{n-1}\}, \text{ one has}$$

$$(3.11) \qquad P_{(u,0)}^\Delta [V_n = u_n, A | S_k < \infty] = p_V(u_{n-1}, u_n) P_{(u,0)}^\Delta [A | S_k < \infty].$$

3. Upper bound

Using the strong Markov property at time S_n, one has

(3.12) $P^\Delta_{(u,0)}[V_n = u_n, A | S_k < \infty] =$

$$\overset{(\text{Markov})}{=} \frac{E^\Delta_{(u,0)}\left[V_n = u_n, A, S_n < \infty, P^\Delta_{(V_n,0)}[S_{k-n} < \infty]\right]}{P^\Delta_0[S_k < \infty]}$$

$$= \frac{P^\Delta_0[S_{k-n} < \infty]}{P^\Delta_0[S_k < \infty]} P^\Delta_{(u,0)}[V_n = u_n, A, S_n < \infty],$$

where in the last step we have used that the distribution of S_{k-n} is translation invariant in the $\mathbb{T}^d_N$ direction. Applying the strong Markov property at time S_{n-1} to the last probability in this expression, we find

$$P^\Delta_{(u,0)}[V_n = u_n, A, S_n < \infty] =$$

$$\overset{(\text{Markov})}{=} E^\Delta_{(u,0)}\left[A, S_{n-1} < \infty, P^\Delta_{(V_{n-1},0)}[V_1 = u_n, S_1 < \infty]\right]$$

$$\overset{((3.8),(3.10))}{=} p_V(u_{n-1}, u_n) E^\Delta_{(u,0)}\left[A, S_{n-1} < \infty, P^\Delta_{(u_{n-1},0)}[S_1 < \infty]\right]$$

$$\overset{((3.10),(\text{Markov}))}{=} p_V(u_{n-1}, u_n) P^\Delta_{(u,0)}[A, S_n < \infty].$$

Substituting this last expression into (3.12), and noting that (once more by the strong Markov property)

$$P^\Delta_0[S_{k-n} < \infty] P^\Delta_{(u,0)}[A, S_n < \infty]$$
$$= E^\Delta_{(u,0)}\left[A, S_n < \infty, P^\Delta_{(u_{n-1},0)}[S_{k-n} < \infty]\right]$$
$$\overset{(\text{Markov})}{=} P^\Delta_{(u,0)}[A, S_k < \infty],$$

we obtain (3.11) and finish the proof of Lemma 3.3. □

With the notation of Lemma 3.2, we recall that $\lambda_m(V)$ and $\lambda_m(Y)$, $m = 1, \ldots, N^d$ stand for the decreasingly ordered eigenvalues of the transition matrices $(p_V(u, u'))_{u,u' \in \mathbb{T}^d_N}$ and $(p_Y(u, u'))_{u,u' \in \mathbb{T}^d_N}$ of V and Y respectively. The following proposition shows how these two sets of eigenvalues are related.

PROPOSITION 3.4. ($d \geq 1$, $\Delta \in [0, 1)$)

(3.13) $\quad \lambda_m(V) = E^\Delta_0\left[\lambda_m(Y)^{\rho^Y_{S_1}} \Big| S_1 < \infty\right], \quad 1 \leq m \leq N^d.$

PROOF. From (3.9), we know that, for $u, u' \in \mathbb{T}^d_N$,

$$p_V(u, u') = E^\Delta_0\left[p_Y^{\rho^Y_{S_1}}(u, u') \Big| S_1 < \infty\right].$$

For any eigenvalue/eigenvector pair $(\lambda_m(Y), v_m)$, we infer that

$$(p_V(u, u'))_{u,u'} v_m = E^\Delta_0\left[(p_Y(u, u'))^{\rho^Y_{S_1}}_{u,u'} v_m \Big| S_1 < \infty\right]$$
$$= E^\Delta_0\left[\lambda_m(Y)^{\rho^Y_{S_1}} v_m \Big| S_1 < \infty\right] = E^\Delta_0\left[\lambda_m(Y)^{\rho^Y_{S_1}} \Big| S_1 < \infty\right] v_m.$$

38   1. Disconnection of a discrete cylinder by a biased random walk

Hence, $(p_V(u,u'))_{u,u' \in \mathbb{T}_N^d}$ has the same eigenvectors as $(p_Y(u,u'))_{u,u' \in \mathbb{T}_N^d}$ and the corresponding eigenvalues are indeed given by (3.13). □

We can thus relate the quantity $u(V)$ to $u(Y)$ (cf. (3.6)), which is well known from Aldous and Fill [3]:

PROPOSITION 3.5. $(d \geq 2, N \geq 1)$

(3.14) $\qquad u(Y) \leq cN^2 \log N \qquad (d=2),$

(3.15) $\qquad u(Y) \leq cN^d \qquad (d \geq 3).$

(We refer to the end of the introduction for our convention concerning constants.)

PROOF. The proof is contained in [3]: By the eigentime identity from Chapter 3, Proposition 13, $u(Y)$ is equal to the average hitting time (cf. Chapter 4, p. 1, for the definition), for which the estimates hold by Proposition 8 in Chapter 13. □

As a consequence, we now obtain our desired estimate on $C_{\mathbb{T}_N^d}^V$ by an application of Lemma 3.2:

LEMMA 3.6. $(d \geq 2, N \geq 2, u \in \mathbb{T}_N^d)$ For any $k \geq [c_1 N^d (\log N)^2]$, one has

(3.16) $\qquad \sup_{\Delta \in [0,1)} P_{(u,0)}^\Delta \left[C_{\mathbb{T}_N^d}^V \geq [c_1 N^d (\log N)^2] \Big| S_k < \infty \right] \leq \dfrac{1}{N^{10}}.$

PROOF. We fix any $\Delta \in [0,1)$ and consider the canonical Markov chain $(W_n)_{n \geq 0}$, with state-space $\mathbb{T}_N^d$, starting point u and with the same transition probability as $(V_n)_{n=0}^k$ under $P_{(u,0)}^\Delta[\cdot | S_k < \infty]$, i.e. $p_W(\cdot,\cdot) = p_V(\cdot,\cdot)$. By Lemma 3.3, $(W_n)_{n \geq 0}$ then satisfies the assumptions of Lemma 3.2. Moreover, $(W_n)_{n=0}^k$ has the same distribution as $(V_n)_{n=0}^k$ under $P_{(u,0)}^\Delta[\cdot | S_k < \infty]$. With the help of Lemma 3.2, we see that, for $k \geq [cN^d (\log N)^2]$,

(3.17) $\qquad P_{(u,0)}^\Delta \left[C_{\mathbb{T}_N^d}^V \geq [cN^d (\log N)^2] \Big| S_k < \infty \right]$
$\qquad \qquad = P_u \left[C_{\mathbb{T}_N^d}^W \geq [cN^d (\log N)^2] \right]$
$\qquad \qquad \overset{(3.5)}{\leq} N^d \exp \left\{ - \left[\dfrac{[cN^d (\log N)^2]}{4eu(W)} \right] \right\}.$

Since V and W have the same transition probability, we have $u(W) = u(V)$, so once we show that

(3.18) $\qquad u(V) = \sum_{m=2}^{N^d} \dfrac{1}{1-\lambda_m(V)} \leq c(N^d + u(Y)),$

the proof of (3.16) will be complete with (3.14), (3.15), (3.17) by choosing $c = c_1$ a large enough constant and noting that the right-hand side of (3.17) does not depend on Δ. We use the expression for

3. Upper bound

$\lambda_m(V)$ of (3.13) and distinguish the two cases $0 < \lambda_m(Y) < 1$ and $-1 \leq \lambda_m(Y) \leq 0$. If $0 < \lambda_m(Y) < 1$, then

$$\lambda_m(V) = E_0^\Delta \left[\lambda_m(Y)^{\rho_{S_1}^Y} \big| S_1 < \infty \right] \leq p + \lambda_m(Y)(1-p),$$

where

(3.19)
$$\begin{aligned} p &= P_0^\Delta[\rho_{S_1}^Y = 0 | S_1 < \infty] \\ &\leq 1 - P_0^\Delta[\rho_{S_1}^Y = 1, S_1 < \infty] \\ &\leq 1 - P_0^\Delta[X_1 \in \mathbb{T}_N^d \times \{0\}] = \frac{1}{d+1}. \end{aligned}$$

We deduce that for $0 < \lambda_m(Y) < 1$,

(3.20)
$$\frac{1}{1 - \lambda_m(V)} \leq \frac{1}{1-p} \frac{1}{1 - \lambda_m(Y)} = \frac{d+1}{d} \frac{1}{1 - \lambda_m(Y)}.$$

If, on the other hand, $-1 \leq \lambda_m(Y) \leq 0$, then $\lambda_m(Y)^n$ is non-negative only for even $n \geq 1$ and not larger than 1 for all $n \geq 1$, so in particular

$$\begin{aligned} \lambda_m(V) &\leq P_0^\Delta \left[\rho_{S_1}^Y \geq 2 \big| \bar{S}_1 < \infty \right] \\ &\leq 1 - P_0^\Delta[\rho_{S_1}^Y = 1, S_1 < \infty] \overset{(3.19)}{\leq} \frac{1}{d+1}, \end{aligned}$$

and hence

(3.21)
$$\frac{1}{1 - \lambda_m(V)} \leq \frac{d+1}{d}.$$

The estimates (3.20) and (3.21) together yield (3.18), so the proof of Lemma 3.6 is completed. $\square$

In view of (3.16), we still need an upper bound on the amount of time it takes for the corresponding $[c_1 N^d (\log N)^2]$ returns to occur. For simplicity of notation, we set

(3.22) $$a_N = N^d (\log N)^2,$$

and we treat the cases $\alpha > 1$ and $\alpha \leq 1$ in Theorem 3.1 separately.

Case $\alpha > 1$. We observe that

(3.23) $$P_0^{N^{-d\alpha}} \left[T_N > a_N^2 (\log N)^\epsilon \right]$$
$$\leq P_0^{N^{-d\alpha}} \left[T_N > S_{[c_1 a_N]} \right] + P_0^{N^{-d\alpha}} \left[S_{[c_1 a_N]} > a_N^2 (\log N)^\epsilon \right].$$

By Lemma 3.6 one has

$$\begin{aligned} P_0^{N^{-d\alpha}} \left[T_N > S_{[c_1 a_N]} \right] &\leq P_0^{N^{-d\alpha}} \left[C_{\mathbb{T}_N^d \times \{0\}}^X > S_{[c_1 a_N]} \right] \\ &= P_0^{N^{-d\alpha}} \left[C_{\mathbb{T}_N^d}^V > [c_1 a_N], S_{[c_1 a_N]} < \infty \right] \overset{(3.16)}{\longrightarrow} 0. \end{aligned}$$

In view of (3.23), the proof of (3.1) will thus be complete once it is shown that

(3.24) $$P_0^{N^{-d\alpha}} \left[S_{[c_1 a_N]} \leq a_N^2 (\log N)^\epsilon \right] \overset{N \to \infty}{\longrightarrow} 1.$$

40   1. Disconnection of a discrete cylinder by a biased random walk

Let us first remove the drift. By (2.22) of Lemma 2.1, applied with $T = S_{[c_1 a_N]}$, $A = \{S_{[c_1 a_N]} \leq a_N^2 (\log N)^\epsilon\}$ and $b = 0$, the probability in (3.24) is bounded from below by

$$(1 - N^{-2d\alpha})^{ca_N^2(\log N)^\epsilon} P_0^0 \left[S_{[c_1 a_N]} \leq a_N^2 (\log N)^\epsilon \right],$$

and since $\alpha > 1$, the factor before the probability tends to 1 as $N \to \infty$. It therefore only remains to show that

$$(3.25) \qquad P_0^0 \left[S_{[c_1 a_N]} \leq a_N^2 (\log N)^\epsilon \right] \xrightarrow{N \to \infty} 1.$$

Observe that when the random walk X, started in $\mathbb{T}_N^d \times \{0\}$, has entered the set $\mathbb{T}_N^d \times \{1\}$ and returned to $\mathbb{T}_N^d \times \{0\}$, then the time $\tilde{H}_{\mathbb{T} \times \{0\}}$ must have passed. In other words, one has

$$(3.26) \qquad \tilde{H}_{\mathbb{T} \times \{0\}}^X \leq H_{\{0\}}^{\pi_\mathbb{Z}(X)} \circ \theta_{H_{\{1\}}^{\pi_\mathbb{Z}(X)}} + H_{\{1\}}^{\pi_\mathbb{Z}(X)}, \quad P_0^0\text{-a.s.}$$

With (3.3) and the strong Markov property applied at the successive entrance times of $\mathbb{T}_N^d \times \{1\}$ and $\mathbb{T}_N^d \times \{0\}$, one deduces that $S_{[c_1 a_N]}$ is stochastically dominated by $H_{2[c_1 a_N]}^{\pi_\mathbb{Z}(X)}$. In particular, we infer that

$$(3.27) \qquad P_0^0 \left[S_{[c_1 a_N]} \leq a_N^2 (\log N)^\epsilon \right] \geq P_0^0 \left[H_{2[c_1 a_N]}^{\pi_\mathbb{Z}(X)} \leq a_N^2 (\log N)^\epsilon \right].$$

From (2.11), we obtain further that

$$(3.28) \quad P_0^0 \left[H_{2[c_1 a_N]}^{\pi_\mathbb{Z}(X)} \leq a_N^2 (\log N)^\epsilon \right]$$
$$\geq P_0^0 \left[H_{2[c_1 a_N]}^Z \leq \frac{a_N^2 (\log N)^\epsilon}{2(d+1)}, \rho_{[a_N^2(\log N)^\epsilon]}^Z \geq \frac{a_N^2 (\log N)^\epsilon}{2(d+1)} \right].$$

The probability of the first event on the right-hand side of (3.28) tends to 1 as $N \to \infty$ by the invariance principle for the one-dimensional simple random walk Z, cf. (2.15). Since by (2.16), $\rho_{[a_N^2(\log N)^\epsilon]}^Z$ is distributed a sum of $[a_N^2 (\log N)^\epsilon]$ iid random variables of expectation $1/(d+1)$, the probability of the second event on the right-hand side of (3.28) tends to 1 by the law of large numbers. Hence, (3.25) follows from (3.27) and (3.28) and the proof of Theorem 3.1 for $\alpha > 1$ is complete.

Case $\alpha \leq 1$. We claim that in order to prove (3.2), it suffices to show that for some constant $c_2(c_1) > 0$ and $N \geq c(c_1)$, with a_N defined in (3.22),

$$(3.29) \qquad P_0^{N^{-d\alpha}} \left[X([0, N^{3d})) \supseteq \mathbb{T}_N^d \times \{0\} \right] \geq e^{-c_2 N^{-d\alpha} a_N}$$

(recall our convention concerning constants from the end of the introduction). Indeed, suppose that (3.29) holds true. Then observe that, on the event $\{T \geq e^{c_0 N^{-d\alpha} a_N}\}$, X does not cover $\mathbb{T}_N^d \times \{Z_{n N^{3d}}\}$ during the time interval $[n N^{3d}, (n+1) N^{3d})$ for $0 \leq n \leq \left[N^{-3d} e^{c_0 N^{-d\alpha} a_N} \right] - 1$,

3. Upper bound

$n \geq 1$, for covering of a slice of E results in the disconnection of E. We thus apply the simple Markov property inductively at times

$$\left\{nN^{3d} : n = \left[N^{-3d}e^{c_0 N^{-d\alpha} a_N}\right] - 1, \ldots, 2, 1\right\},$$

and obtain

$$P_0^{N-d\alpha}\left[T_N \geq e^{c_0 N^{-d\alpha} a_N}\right] \leq$$

$$\leq P_0^{N-d\alpha}\left[\bigcap_{n=0}^{[N^{-3d}e^{c_0 N^{-d\alpha} a_N}]-1} \theta_{nN^{3d}}^{-1}\{X([0, N^{3d})) \not\supseteq \mathbb{T}_N^d \times \{Z_{nN^{3d}}\}\}\right]$$

$$\stackrel{\text{(Markov, transl. inv.)}}{=} P_0^{N-d\alpha}\left[X([0, N^{3d})) \not\supseteq \mathbb{T}_N^d \times \{0\}\right]^{[N^{-3d}e^{(c_0 N^{-d\alpha} a_N)}]}$$

$$\stackrel{(3.29)}{\leq} \exp\left\{-ce^{(c_0 - c_2)N^{-d\alpha} a_N} N^{-3d}\right\}$$

$$\stackrel{(3.22)}{=} \exp\left\{-ce^{(c_0 - c_2)N^{d(1-\alpha)}(\log N)^2} N^{-3d}\right\},$$

so the proof of (3.2) is complete by the fact that $\alpha \leq 1$, provided we choose $c_0 > c_2(c_1)$. It thus remains to establish the estimate (3.29). To this end, we observe that

(3.30) $\quad P_0^{N-d\alpha}\left[X([0, N^{3d})) \supseteq \mathbb{T}_N^d \times \{0\}\right]$
$\quad \geq P_0^{N-d\alpha}\left[X([0, \infty)) \supseteq \mathbb{T}_N^d \times \{0\}\right]$
$\quad - P_0^{N-d\alpha}\left[X([N^{3d}, \infty)) \cap \mathbb{T}_N^d \times \{0\} \neq \emptyset\right].$

Standard large deviation estimates allow us to bound the second probability on the right-hand side:

$$P_0^{N-d\alpha}\left[X([N^{3d}, \infty)) \cap \mathbb{T}_N^d \times \{0\} \neq \emptyset\right]$$
$$= P_0^{N-d\alpha}\left[\text{for some } n \geq N^{3d}, \pi_\mathbb{Z}(X_n) = 0\right]$$
$$\leq \sum_{n \geq N^{3d}} P_0^{N-d\alpha}\left[\pi_\mathbb{Z}(X_n) - N^{-d\alpha}\frac{n}{d+1} < -\frac{1}{2}N^{-d\alpha}\frac{n}{d+1}\right].$$

Now observe that $(\pi_\mathbb{Z}(X_n) - \Delta n/(d+1))_{n \geq 0}$ is a P_0^Δ-martingale with increments bounded by $1 + \Delta/(d+1) \leq 2$. By Azuma's inequality (see, for instance, [4], p. 85), the expression in the last sum is therefore bounded from above by $\exp\{-cN^{-2d\alpha}n\}$. This yields

$$P_0^{N-d\alpha}\left[X([N^{3d}, \infty)) \cap \mathbb{T}_N^d \times \{0\} \neq \emptyset\right] \leq \exp\left\{-cN^{3d}N^{-2d\alpha}\right\}$$
$$\stackrel{(\alpha \leq 1)}{\leq} \exp\{-cN^d\}.$$

Inserting this last estimate into (3.30), we see that in fact (3.29) will follow from

(3.31) $\quad P_0^{N-d\alpha}\left[X([0, \infty)) \supseteq \mathbb{T}_N^d \times \{0\}\right] \geq ce^{-\frac{1}{2}c_2 N^{-d\alpha} a_N}.$

42   1. Disconnection of a discrete cylinder by a biased random walk

By (3.16), we have

(3.32) $P_0^{N-d\alpha}\left[X([0,\infty)) \supseteq \mathbb{T}_N^d \times \{0\}\right] \geq$

$\geq P_0^{N-d\alpha}\left[S_{[c_1a_N]} < \infty, X([0, S_{[c_1a_N]})) \supseteq \mathbb{T}_N^d \times \{0\}\right]$

$= P_0^{N-d\alpha}\left[S_{[c_1a_N]} < \infty\right] P_0^{N-d\alpha}\left[C_{\mathbb{T}_N^d}^V \leq [c_1a_N] \Big| S_{[c_1a_N]} < \infty\right]$

$\overset{((3.16),(3.22))}{\geq} cP_0^{N-d\alpha}\left[S_{[c_1a_N]} < \infty\right].$

We use again the estimate (3.26) and deduce that $S_{[c_1a_N]}$ is stochastically dominated by $H_{-2[c_1a_N]}^{\pi_{\mathbb{Z}}(X)}$. In particular, we have

(3.33) $P_0^{N-d\alpha}\left[S_{[c_1a_N]} < \infty\right] \geq P_0^{N-d\alpha}\left[H_{-2[c_1a_N]}^{\pi_{\mathbb{Z}}(X)} < \infty\right]$

$= P_0^{N-d\alpha}\left[H_{-2[c_1a_N]}^Z < \infty\right]$

$= \left(\frac{1 - N^{-d\alpha}}{1 + N^{-d\alpha}}\right)^{2[c_1a_N]} \geq e^{-c(c_1)N^{-d\alpha}a_N},$

using a standard estimate on one-dimensional biased random walk for the last line (see, for example, Durrett [15] Chapter 4, Example 7.1 (c), p. 272). The estimates (3.32) and (3.33) together show (3.31) for a suitably chosen constant $c_2(c_1) > 0$. Hence, the proof of (3.2) and thus of Theorem 3.1 is complete. □

4. Lower bounds: Reduction to large deviations

The goal of this section is to prove Theorem 1.2 reducing the problem of finding a lower bound on T_N to a large deviation estimate of the form (1.10). As a preliminary step towards this reduction, we prove the following geometric lemma in the spirit of Dembo and Sznitman [12], where we refer to (1.6) for our notion of κ-disconnection:

LEMMA 4.1. $(d \geq 1\ \alpha > 0,\ \kappa \in (0, \frac{1}{2}))$ There is a constant $c(\alpha, \kappa)$ such that for all $N \geq c(\alpha, \kappa)$, whenever $K \subseteq E$ disconnects E, there is an $x_* \in E$ such that K κ-disconnects $x_* + B(\alpha)$, cf. (1.7). (We refer to the end of the introduction for our convention concerning constants.)

PROOF. We follow the argument contained in the proof of Lemma 2.4 in Dembo and Sznitman [12]. Assuming that K disconnects E, we refer as Top to the connected component of $E \setminus K$ containing $\mathbb{T}_N^d \times [M, \infty)$ for large $M \geq 1$. We can then define the function

$$t: E \longrightarrow \mathbb{R}_+$$
$$x \mapsto \frac{|Top \cap (x + B(\alpha))|}{|B(\alpha)|}.$$

The function t takes the value 0 for $x = (u, v) \in E$ with $v \in \mathbb{Z}$ a large negative number and the value 1 for v a large positive number.

4. Lower bounds: Reduction to large deviations   43

Moreover, for $x = (u,v)$, $x' = (u,v') \in E$ such that $|v - v'| = 1$ we have (with Δ denoting symmetric difference),

$$|t(x) - t(x')| \leq \frac{|(x + B(\alpha))\Delta(x' + B(\alpha))|}{|B(\alpha)|} \leq \frac{cN^d}{N^{d+d\alpha\wedge 1}} = \frac{c}{N^{d\alpha\wedge 1}}.$$

Using these last two observations on t, we see that, for $N \geq c(\alpha, \kappa)$, there is at least one $x_* \in E$ satisfying

$$\left|t(x_*) - \frac{1}{2}\right| \leq \frac{c}{N^{d\alpha\wedge 1}} \leq \frac{1}{2} - \kappa,$$

which can be restated as

(4.1) $\quad \kappa|B(\alpha)| \leq |Top \cap (x_* + B(\alpha))| \leq (1 - \kappa)|B(\alpha)|.$

If we set $I = Top \cap (x_* + B(\alpha))$, then $\partial_{(x_* + B(\alpha))}(I) \subseteq K$ (since K disconnects E), so that the proof is complete with (4.1). $\square$

PROOF OF THEOREM 1.2. We claim that it suffices to prove the following two estimates on $P_0^{N^{-d\alpha}}[T_N \leq t]$, valid for any $t \geq 1$, $\xi \in (0, f(\alpha, \beta))$ (for $\alpha, \beta > 0$ and f as in (1.10)) and $N \geq c(\alpha, \beta, \xi)$:

(4.2) $\quad P_0^{N^{-d\alpha}}[T_N \leq t] \leq cN^d(t + N)\left(e^{-N^\xi} + e^{-c'N^{\beta + (d\alpha \wedge 1)}t^{-\frac{1}{2}}}\right).$

and

(4.3) $\quad P_0^{N^{-d\alpha}}[T_N \leq t] \leq cN^d(t + N)\left(e^{-N^\xi} + e^{-c'N^{\beta - (d\alpha - 1)_+}}\right)$

Indeed, suppose that (4.2) and (4.3) both hold. In order to deduce (1.11), we then choose any $\alpha > 1$, $0 < \epsilon < 2d$ such that $\beta = d - 1 - \frac{\epsilon}{4} > 0$ (note $d \geq 2$) and $\xi \in (0, f(\alpha, \beta))$ (which is possible by the assumption on f). With $t = N^{2d-\epsilon}$, (4.2) then yields, for $N \geq c(\alpha, \beta, \xi, \epsilon)$,

$$P_0^{N^{-d\alpha}}[T_N \leq N^{2d-\epsilon}] \leq cN^{3d-\epsilon}\left(e^{-N^\xi} + e^{-c'N^{\frac{\epsilon}{4}}}\right),$$

and hence shows (1.11).

On the other hand, choosing $t = \exp\{N^\mu\}$, $\mu > 0$, in (4.3), we have, for any $\alpha, \beta > 0$, $\xi \in (0, f(\alpha, \beta))$ and $N \geq c(\alpha, \beta, \xi, \mu)$,

(4.4) $\quad P_0^{N^{-d\alpha}}[T_N \leq \exp\{N^\mu\}]$
$$\leq cN^d \left(\exp\{N^\mu - N^\xi\} + \exp\{N^\mu - c'N^{\beta - (d\alpha - 1)_+}\}\right).$$

The right-hand side of (4.4) tends to 0 as $N \to \infty$ for α, β, ξ as above, provided $\beta > (d\alpha - 1)_+$ and $\mu < \xi \wedge (\beta - (d\alpha - 1)_+)$. We thus obtain (1.12) by optimizing over β and ξ in (4.4).

It therefore remains to establish (4.2) and (4.3). To this end, we apply the geometric Lemma 4.1, noting that, up to time t, only sets $(u,v) + B(\alpha)$ (in the notation of (1.7)) with $|v| \leq t + N^{d\alpha \wedge 1}$ can be entered by the discrete-time random walk, and thus deduce that, for $N \geq c(\alpha)$,

(4.5) $\quad P_0^{N-d\alpha}[T_N \leq t] \leq cN^d(t+N) \times$

$$\sup_{x \in E} P_0^{N-d\alpha}\left[X([0,[t]]) \tfrac{1}{3}\text{-disconnects } x + B(\alpha)\right].$$

For the first return time $\mathcal{R}_1^x$, defined as $\mathcal{R}_1^x = H_{S_{2[N^{d\alpha}\wedge 1]}}^X$ (cf. (2.17)), one has

$$\left\{X([0,[t]]) \tfrac{1}{3}\text{-disconnects } x + B(\alpha)\right\}$$
$$\subseteq \theta_{\mathcal{R}_1^x}^{-1}\left\{X([0,[t]]) \tfrac{1}{3}\text{-disconnects } x + B(\alpha)\right\}.$$

Applying the strong Markov property at time $\mathcal{R}_1^x$ and using translation invariance, we thus obtain that (cf. (1.8))

$$P_0^{N-d\alpha}\left[X([0,[t]]) \tfrac{1}{3}\text{-disconnects } x + B(\alpha)\right]$$
$$\leq \sup_{x \in S_{2[N^{d\alpha}\wedge 1]}} P_x^{N-d\alpha}\left[X([0,[t]]) \tfrac{1}{3}\text{-disconnects } B(\alpha)\right]$$
$$= \sup_{x \in S_{2[N^{d\alpha}\wedge 1]}} P_x^{N-d\alpha}\left[U_{B(\alpha)} \leq t\right].$$

Inserted into (4.5), this yields

(4.6) $\quad P_0^{N-d\alpha}[T_N \leq t] \leq cN^d(t+N) \sup_{x \in S_{2[N^{d\alpha}\wedge 1]}} P_x^{N-d\alpha}\left[U_{B(\alpha)} \leq t\right].$

We then observe that, for any $x \in S_{2[N^{d\alpha}\wedge 1]}$,

(4.7) $\quad P_x^{N-d\alpha}\left[U_{B(\alpha)} \leq t\right]$
$$\leq P_x^{N-d\alpha}\left[U_{B(\alpha)} < \mathcal{D}_{[N^\beta]}\right] + P_x^{N-d\alpha}\left[\mathcal{R}_{[N^\beta]} \leq U_{B(\alpha)} \leq t\right]$$
$$\stackrel{(\text{def.})}{=} P_1 + P_2.$$

By definition of $U_{B(\alpha)}$ we know that, on the event $\{U_{B(\alpha)} < \infty\}$, $\pi_{\mathbb{Z}}(X_{U_{B(\alpha)}} - x) \leq c[N^{d\alpha\wedge 1}]$, $P_x^{N-d\alpha}$-a.s., for $x \in S_{2[N^{d\alpha}\wedge 1]}$. We can thus apply (2.23) of Lemma 2.1 with $A = \{U_{B(\alpha)} < \mathcal{D}_{[N^\beta]}\}$, $T = U_{B(\alpha)}$ and $b' = c[N^{d\alpha\wedge 1}]$ and obtain, for P_1 in (4.7),

(4.8) $\quad P_1 \stackrel{(2.23)}{\leq} (1+N^{-d\alpha})^{c[N^{d\alpha\wedge 1}]} P_x^0\left[U_{B(\alpha)} < \mathcal{D}_{[N^\beta]}\right]$
$$\leq cP_x^0\left[U_{B(\alpha)} < \mathcal{D}_{[N^\beta]}\right] \stackrel{(1.10)}{\leq} e^{-N^\xi},$$

for any $\xi \in (0, f(\alpha,\beta))$ and all $N \geq c(\alpha,\beta,\xi)$. Turning to P_2 in (4.7), we apply (2.23) of Lemma 2.1 with $A = \{\mathcal{R}_{[N^\beta]} \leq t\}$, $T = \mathcal{R}_{[N^\beta]}$ and

4. Lower bounds: Reduction to large deviations   45

$b' = c[N^{d\alpha \wedge 1}]$, and obtain

$$\begin{aligned}P_2 &\leq P_x^{N-d\alpha}\left[\mathcal{R}_{[N^\beta]} \leq t\right] \\ &\leq (1+N^{-d\alpha})^{c[N^{d\alpha \wedge 1}]} P_x^0\left[\mathcal{R}_{[N^\beta]} \leq t\right] \\ &\leq c P_x^0\left[\mathcal{R}_{[N^\beta]} \leq t\right].\end{aligned}$$

For this last probability, we make the observation that, under P_x^0, $\mathcal{R}_{[N^\beta]} - \mathcal{D}_1$ ($\leq \mathcal{R}_{[N^\beta]}$) is distributed as the sum of at least $[cN^{d\alpha \wedge 1} N^\beta]$ independent random variables, all of which are distributed as the hitting time of 1 for the unbiased simple random walk $\pi_{\mathbb{Z}}(X)$ (cf. (2.1)) starting at the origin with geometric delay of constant parameter $\frac{1}{d+1}$. Applying an elementary estimate on one-dimensional simple random walk for the second inequality (cf. Durrett [15], Chapter 3, (3.4)), we deduce that, for $t \geq 1$,

$$\begin{aligned}P_2 &\leq c P_0^0\left[H_1^{\pi_{\mathbb{Z}}(X)} \leq t\right]^{cN^{\beta+(d\alpha \wedge 1)}} \\ &\leq c\left(1 - c' t^{-\frac{1}{2}}\right)^{c'N^{\beta+(d\alpha \wedge 1)}} \\ &\leq c\exp\left\{-c' N^{\beta+(d\alpha \wedge 1)} t^{-\frac{1}{2}}\right\}.\end{aligned}$$

Together with (4.8), (4.7) and (4.6), this yields (4.2).

In order to obtain (4.3), we use the following different method for estimating P_2 in (4.7): We let A^- be the event that the random walk X first exits $S_{4[N^{d\alpha \wedge 1}]}$ into the negative direction, i.e.

$$A^- = \{\pi_{\mathbb{Z}}(X_{\mathcal{D}_1}) < 0\} \in \mathcal{F}_{\mathcal{D}_1}.$$

One then has

$$\begin{aligned}(4.9)\quad P_2 &\leq \sup_{x \in S_{2[N^{d\alpha \wedge 1}]}} P_x^{N-d\alpha}\left[\mathcal{R}_{[N^\beta]} < \infty\right] \\ &= \sup_{x \in S_{2[N^{d\alpha \wedge 1}]}} \left(P_x^{N-d\alpha}\left[\mathcal{R}_{[N^\beta]} < \infty, A^-\right] \right.\\ &\qquad\qquad \left.+ P_x^{N-d\alpha}\left[\mathcal{R}_{[N^\beta]} < \infty, (A^-)^c\right]\right).\end{aligned}$$

We now apply the strong Markov property at the times $\mathcal{D}_1$ and $\mathcal{R}_2$ and use translation invariance to infer from (4.9) that, for $N^\beta \geq 2$,

$$(4.10)\quad P_2 \leq \sup_{x \in S_{2[N^{d\alpha \wedge 1}]}} P_x^{N-d\alpha}\left[\mathcal{R}_{[N^\beta]-1} < \infty\right] \times \sup_{x \in S_{2[N^{d\alpha \wedge 1}]}}$$
$$\left(P_x^{N-d\alpha}[A^-] + P_x^{N-d\alpha}[(A^-)^c] P_0^{N-d\alpha}\left[H_{-c[N^{d\alpha \wedge 1}]}^{\pi_{\mathbb{Z}}(X)} < \infty\right]\right).$$

Next, we apply the estimate (2.23) of Lemma 2.1 with $T = \mathcal{D}_1$, $A = A^-$ and $b' = -2[N^{d\alpha \wedge 1}]$, then the invariance principle for one-dimensional

46   1. Disconnection of a discrete cylinder by a biased random walk

simple random walk, and obtain, for any $x \in S_{2[N^{d\alpha \wedge 1}]}$,

(4.11) $P_x^{N^{-d\alpha}}[A^-] \overset{(2.23)}{\leq} P_x^0[A^-] \overset{(\text{inv. princ.})}{\leq} (1 - c_3), \quad c_3 > 0.$

Moreover, since the projection $\pi_{\mathbb{Z}}(X)$ of X on $\mathbb{Z}$ is a one-dimensional random walk with drift $\frac{N^{-d\alpha}}{d+1}$ and geometric delay of constant parameter $\frac{1}{d+1}$, standard estimates on one-dimensional biased random walk imply

(4.12) $P_0^{N^{-d\alpha}}\left[H_{-c[N^{d\alpha \wedge 1}]}^{\pi_{\mathbb{Z}}(X)} < \infty\right] \leq \left(\frac{1 - N^{-d\alpha}(d+1)^{-1}}{1 + N^{-d\alpha}(d+1)^{-1}}\right)^{c[N^{d\alpha \wedge 1}]}$
$\leq e^{-cN^{-d\alpha}[N^{d\alpha \wedge 1}]}.$

Inserting (4.11) and (4.12) into (4.10) and using induction, we deduce

(4.13) $P_2 \leq \left(1 - c_3 + c_3 e^{-cN^{-d\alpha}[N^{d\alpha \wedge 1}]}\right)^{[N^\beta]-1}.$

Note that $N^{-d\alpha}[N^{d\alpha \wedge 1}] \leq 1$. If $d\alpha > 1$, then the right-hand side of (4.13) is bounded from above by $(1 - cN^{-d\alpha}[N^{d\alpha \wedge 1}])^{[N^\beta]-1} \leq e^{-cN^{-d\alpha}[N^{d\alpha \wedge 1}]N^\beta}$, while if $d\alpha \leq 1$, the right-hand side of (4.13) is bounded by e^{-cN^β}. In any case, we infer from (4.13) that

$$P_2 \leq e^{-cN^{-d\alpha}N^{\beta+(d\alpha \wedge 1)}} = e^{-cN^{\beta-(d\alpha-1)_+}}.$$

Together with (4.8), (4.7) and (4.6) this yields (4.3) and completes the proof of Theorem 1.2.   □

5. More geometric lemmas

The purpose of this section is to prove several geometric lemmas needed for the derivation of the large deviation estimate (1.10) in Theorem 1.2. The general purpose of these geometric results is to impose restrictions on a set K $\frac{1}{3}$-disconnecting $B(\alpha)$. This will enable us to obtain an upper bound on the probability appearing in (1.10), when choosing $K = X([0, \mathcal{D}_{[N^\beta]}])$.

Throughout this and the next section, we consider the scales L and l, defined as

(5.1) $l = [N^\gamma], \ L = [N^{\gamma'}], \quad \text{for } 0 < \gamma < \gamma' \wedge d\alpha, \ 0 < \gamma' < 1.$

The crucial geometric estimates come in Lemma 5.3 and its modification Lemma 5.4. These geometric results, in the spirit of Dembo and Sznitman [12], require as key ingredient an isoperimetric inequality of Deuschel and Pisztora [14], see Lemma 5.2. In rough terms, Lemmas 5.3 and 5.5 show that for any set K disconnecting $C(L)$ or $B(\alpha)$ for $d\alpha < 1$ (cf. (1.7), (2.7)), one can find a whole "surface" of subcubes of $C(L)$ or $B(\alpha)$ such that the set K occupies a "surface" of points inside every one of these subcubes. More precisely, it is shown that there exist subcubes $(C_x(l))_{x \in \mathcal{E}}$ (cf. (2.8)) of $C(L)$ or of $B(\alpha)$ with the

5. More geometric lemmas

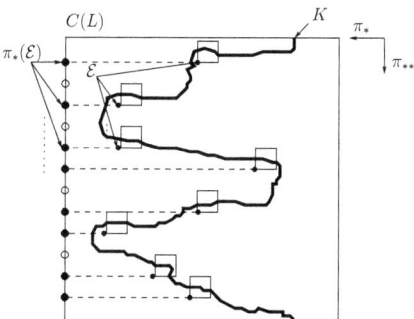

Figure 3: An illustration of the crucial geometric Lemma 5.3. The figure shows the set $C(L)$, disconnected by $K \subseteq C(L)$. The small boxes are the collection of subcubes $(C_x(l))_{x \in \mathcal{E}}$. The circles on the left are the points on the projected subgrid of side-length l, a large number of which (the filled ones) are occupied by the projected set $\pi_*(\mathcal{E})$ of base-points $\mathcal{E}$ (cf. (5.12)), (5.13)). In every subcube, the set K occupies a surface of a significant number of points, in the sense of (5.14).

following properties: for one of the projections π_* on the d-dimensional hyperplanes, the projected set of base-points $\pi_*(\mathcal{E})$ is arranged on a subgrid of side-length l and is substantially large. In the case of $C(L)$, this set of points occupies at least a constant fraction of the volume of the projected subgrid of $C(L)$. Moreover, for one of the projections π_{**} (possibly different from π_*), the π_{**}-projection of the disconnecting set K intersected with any sub-cube $C_x(l)$, $x \in \mathcal{E}$, contains at least cl^d points, i.e. at least a constant fraction of the volume of $\pi_{**}(C_x(l))$ (see Figure 3 for an illustration of the idea).

The first lemma in this section allows to propagate disconnection of the $|.|_\infty$-ball $B_\infty(0, [N/4])$ to a smaller scale of size L, in the sense that, for any set K $\frac{1}{3}$-disconnecting $B_\infty(0, [N/4])$, one can find a subbox $C_{x_*}(L)$ of $B_\infty(0, [N/4])$ which is $\frac{1}{4}$-disconnected by K (cf. (1.6)). This result will prove useful for the case $B(\alpha) = B_\infty(0, [N/4])$ (i.e. if $d\alpha \geq 1$), where we use an upper bound on the number of excursions between $C_{x_*}(L)$ and $\left(C_{x_*}(L)^{(L)}\right)^c$ performed by the random walk X until time $\mathcal{D}_{[N^\beta]}$. We refer to the end of the introduction for our convention concerning constants.

LEMMA 5.1. ($d \geq 1$, $\gamma' \in (0,1)$, $L = [N^{\gamma'}]$, $N \geq 1$) There is a constant $c(\gamma') > 0$ such that for all $N \geq c(\gamma')$, whenever $K \subseteq$

48   1. Disconnection of a discrete cylinder by a biased random walk

$B_\infty(0, [N/4])$ $\frac{1}{3}$-disconnects $B_\infty(0, [N/4])$, there is an $x_* \in B_\infty(0, [N/4])$ such that K $\frac{1}{4}$-disconnects $C_{x_*}(L) \subseteq B_\infty(0, [N/4])$.

PROOF. Since K $\frac{1}{3}$-disconnects $B_\infty(0, [N/4])$, cf. (1.6), there is a set $I \subseteq B_\infty(0, [N/4])$ satisfying

$$\frac{1}{3}|B_\infty(0, [N/4])| \leq |I| \leq \frac{2}{3}|B_\infty(0, [N/4])|$$

and $\partial_{B_\infty(0,[N/4])}(I) \subseteq K$. We want to find a point $x_* \in E$ such that $C_{x_*}(L) \subseteq B_\infty(0, [N/4])$ and

(5.2) $$\frac{1}{4}|C(L)| \leq |C_{x_*}(L) \cap I| \leq \frac{3}{4}|C(L)|.$$

To this end, we introduce the subgrid $\mathcal{B}_L \subseteq B_\infty(0, [N/4])^{(-L)}$ of sidelength L, defined as (cf. (2.3))

(5.3) $\mathcal{B}_L = B_\infty(0, [N/4])^{(-L)} \cap \pi_E\left([-[N/4], [N/4]]^{d+1} \cap L\mathbb{Z}^{d+1}\right).$

The boxes $(C_x(L))_{x \in \mathcal{B}_L}$, see (2.7), (2.8), are disjoint subsets of $B_\infty(0, [N/4])$, and their union covers all but at most $cN^d L$ points of $B_\infty(0, [N/4])$. Hence, we have

(5.4) $$\sum_{x \in \mathcal{B}_L} |I \cap C_x(L)| \leq |I| \leq \sum_{x \in \mathcal{B}_L} |I \cap C_x(L)| + cN^d L, \text{ and}$$

(5.5) $|B_\infty(0, [N/4])| - cN^d L \leq |\mathcal{B}_L||C(L)| \leq |B_\infty(0, [N/4])|.$

We now claim that, for $N \geq c(\gamma')$, there is at least one $x_1 \in \mathcal{B}_L$ such that

(5.6) $$|I \cap C_{x_1}(L)| \leq \frac{3}{4}|C(L)|.$$

Indeed, otherwise it would follow from the definition of I and the left-hand inequalities of (5.4) and (5.5) that

$$\frac{2}{3}|B_\infty(0, [N/4])| \geq |I| \stackrel{(5.4)}{>} \frac{3}{4}|C(L)||\mathcal{B}_L| \stackrel{(5.5)}{\geq} \frac{3}{4}|B_\infty(0, [N/4])| - cN^d L,$$

which due to the definition of L is impossible for $N \geq c(\gamma')$. Similarly, for $N \geq c(\gamma')$, we can find an $x_2 \in \mathcal{B}_L$ such that

(5.7) $$\frac{1}{4}|C(L)| \leq |I \cap C_{x_2}(L)|,$$

for otherwise the right-hand inequalities of (5.4) and (5.5) would yield that $\frac{1}{3}|B_\infty(0, [N/4])| \leq \frac{1}{4}|B_\infty(0, [N/4])| + cN^d L$, thus again leading to a contradiction.

Next, we note that, for any neighbors x and $x' \in B_\infty(0, [N/4])$, one has, with Δ denoting the symmetric difference,

(5.8) $$\left|\frac{|C_x(L) \cap I|}{|C(L)|} - \frac{|C_{x'}(L) \cap I|}{|C(L)|}\right| \leq \frac{|C_x(L) \Delta C_{x'}(L)|}{|C(L)|} \leq \frac{c}{N^{\gamma'}}.$$

5. More geometric lemmas

Since both x_1 and x_2 are in $\mathcal{B}_L \subseteq B_\infty(0, [N/4])^{(-L)}$, we can now choose a nearest-neighbor path $\mathcal{P} = (x_1 = y_1, y_2, \ldots, y_n = x_2)$ from x_1 to x_2 such that $C_{y_i}(L) \subseteq B_\infty(0, [N/4])$ for all $y_i \in \mathcal{P}$. Consider now the first point $x_* = y_{i_*}$ on $\mathcal{P}$ such that $\frac{1}{4}|C(L)| \leq |C_{x_*}(L) \cap I|$, which is well-defined thanks to (5.7). If $x_* = y_1$, then by (5.6), x_* satisfies (5.2). If $x_* \neq y_1$, then by (5.8) and choice of x_*, one also has

$$\frac{1}{4}|C(L)| \leq |C_{x_*}(L) \cap I|$$
$$\stackrel{(5.8)}{\leq} |C_{y_{i_*-1}}(L) \cap I| + \frac{c}{N^{\gamma'}}|C(L)|$$
$$< \left(\frac{1}{4} + \frac{c}{N^{\gamma'}}\right)|C(L)|,$$

hence again (5.2) for $N \geq c(\gamma')$. For $N \geq c(\gamma')$, we have thus found an $x_* \in B_\infty(0, [N/4])$ satisfying $\frac{1}{4}|C(L)| \leq |C_{x_*}(L) \cap I| \leq \frac{3}{4}|C(L)|$ and $C_{x_*}(L) \subseteq B_\infty(0, [N/4])$. Moreover,

$$\partial_{C_{x_*}(L)}(C_{x_*}(L) \cap I) \subseteq \partial_{B_\infty(0,[N/4])}(I) \subseteq K.$$

In other words, K $\frac{1}{4}$-disconnects $C_{x_*}(L) \subseteq B_\infty(0, [N/4])$. $\square$

The following lemma contains the essential ingredients for the proof of the two main geometric lemmas thereafter.

LEMMA 5.2. ($d \geq 1$, $\kappa \in (0,1)$, $M \in \{0, \ldots, N-1\}$, $N \geq 1$) Suppose $A \subseteq [0, M]^{d+1} \subseteq E$. Then there is an $i_0 \in \{1, \ldots, d+1\}$ such that

(5.9) $$|A| \leq |\pi_{i_0}(A)|^{\frac{d+1}{d}}.$$

If A in addition satisfies

(5.10) $$|A| \leq (1-\kappa)(M+1)^{d+1},$$

then there is an $i_1 \in \{1, \ldots, d+1\}$ and a constant $c(\kappa) > 0$ such that (cf. (2.6))

(5.11) $$\left|\pi_{i_1}(\partial_{[0,M]^{d+1}, i_1}(A))\right| \geq c(\kappa)|A|^{\frac{d}{d+1}}.$$

PROOF. The estimate (5.9) follows for instance from a theorem of Loomis and Whitney [21]. The proof of (5.11) can be found in equations (A.3)-(A.6) in Deuschel and Pisztora [14], p. 480. $\square$

We now come to the main geometric lemma, which provides a necessary criterion for disconnection of the box $C(L)$ (cf. (2.7)). A schematic illustration of its content can be found in Figure 3.

LEMMA 5.3. ($d \geq 1$, $0 < \gamma < \gamma' < 1$, $l = [N^\gamma]$, $L = [N^{\gamma'}]$, $N \geq 1$) For all $N \geq c(\gamma, \gamma')$, whenever $K \subseteq C(L)$ $\frac{1}{4}$-disconnects $C(L)$ (cf. (1.6)), then there exists a set $\mathcal{E} \subseteq C(L)^{(-l)}$ (cf. (2.3)) and projections π_* and $\pi_{**} \in \{\pi_1, \ldots, \pi_{d+1}\}$ such that

(5.12) $\qquad \pi_*(\mathcal{E}) \subseteq \pi_* \left(C(L) \cap \pi_E \left([0, L]^{d+1} \cap l\mathbb{Z}^{d+1} \right) \right),$

(5.13) $\qquad |\pi_*(\mathcal{E})| \geq c' \left(\dfrac{L}{l} \right)^d,$ and

(5.14) $\qquad$ for all $x \in \mathcal{E}$: $|\pi_{**}(K \cap C_x(l))| \geq c'' l^d$ (cf. (2.8)).

PROOF. Since K $\tfrac{1}{4}$-disconnects $C(L)$, there exists a set $I \subseteq C(L)$ satisfying $\tfrac{1}{4} L^{d+1} \leq |I| \leq \tfrac{3}{4} L^{d+1}$ and $\partial_{C(L)}(I) \subseteq K$. We introduce here the subgrid $\mathcal{C}_l \subseteq C(L)^{(-l)}$ of side-length l, i.e.

(5.15) $\qquad \mathcal{C}_l = C(L)^{(-l)} \cap \pi_E \left([0, L]^{d+1} \cap l\mathbb{Z}^{d+1} \right),$

with subboxes $C_x(l)$, $x \in \mathcal{C}_l$. The set $\mathcal{A}$ is then defined as the set of all $x \in \mathcal{C}_l$ whose corresponding box $C_x(l)$ is filled up to more than $\tfrac{1}{8}$th by I:

(5.16) $\qquad \mathcal{A} = \left\{ x \in \mathcal{C}_l : |C_x(l) \cap I| > \dfrac{1}{8} l^{d+1} \right\}.$

Since the disjoint union of the boxes $(C_x(l))_{x \in \mathcal{C}_l}$ contains all but at most $cL^d l$ points of $C(L)$, we have

(5.17) $\qquad \dfrac{1}{4} L^{d+1} \leq |I| \leq \dfrac{1}{8} l^{d+1} |\mathcal{C}_l \setminus \mathcal{A}| + l^{d+1} |\mathcal{A}| + cL^d l.$

Using the estimate $|\mathcal{C}_l \setminus \mathcal{A}| \leq |\mathcal{C}_l| \leq \left(\tfrac{L}{l} \right)^{d+1}$ and rearranging, we deduce from (5.17) that

$$\left(\dfrac{1}{8} - c \dfrac{l}{L} \right) \left(\dfrac{L}{l} \right)^{d+1} \leq |\mathcal{A}|,$$

so that for $N \geq c(\gamma, \gamma')$,

(5.18) $\qquad \dfrac{1}{9} |\mathcal{C}_l| \leq \dfrac{1}{9} \left(\dfrac{L}{l} \right)^{d+1} \leq |\mathcal{A}|.$

In order to apply the isoperimetric inequality (5.11) of Lemma 5.2 with $\mathcal{A}$ and $\mathcal{C}_l$ playing the roles of A and $[0, M]^{d+1}$ for $N \geq c(\gamma, \gamma')$, we need to keep $|\mathcal{A}|$ away from $|\mathcal{C}_l|$. We therefore distinguish two cases, as to whether or not

(5.19) $\qquad |\mathcal{A}| \leq c_4 |\mathcal{C}_l|,$ with $c_4^3 = \dfrac{1}{2} \left(1 + \dfrac{4}{5} \right).$

Suppose first that (5.19) holds. Then for $N \geq c(\gamma, \gamma')$, the isoperimetric inequality (5.11), applied on the subgrid $\mathcal{C}_l$, yields an $i \in \{1, \ldots, d+1\}$ such that

(5.20) $\qquad |\pi_i (\partial_{\mathcal{C}_l, i}(\mathcal{A}))| \geq c |\mathcal{A}|^{\frac{d}{d+1}} \stackrel{(5.18)}{\geq} c' \left(\dfrac{L}{l} \right)^d,$

5. More geometric lemmas

where $\partial_{\mathcal{C}_l,i}(\mathcal{A})$ denotes the boundary on the subgrid $\mathcal{C}_l$, defined in analogy with (2.6). In order to construct the set $\mathcal{E}$, we apply the following procedure. Given $w \in \pi_i(\partial_{\mathcal{C}_l,i}(\mathcal{A}))$, we choose an $x' \in \partial_{\mathcal{C}_l,i}(\mathcal{A})$ with $\pi_i(x') = w$. In view of (2.6), at least one of $x' + le_i$ and $x' - le_i$ belongs to $\mathcal{A}$. Without loss of generality, we assume that $x' + le_i \in \mathcal{A}$. We then have $|C_{x'}(l) \cap I| \leq \frac{1}{8}l^{d+1}$ (because $x' \in \mathcal{C}_l \setminus \mathcal{A}$, cf. (5.16)) and $|C_{x'+le_i}(l) \cap I| > \frac{1}{8}l^{d+1}$ (because $x'+le_i \in \mathcal{A}$). Observe that neighboring $x_1, x_2 \in E$ satisfy

$$(5.21) \qquad \left| \frac{|C_{x_1}(l) \cap I|}{l^{d+1}} - \frac{|C_{x_2}(l) \cap I|}{l^{d+1}} \right| \leq \frac{c}{N^\gamma}.$$

Now consider the first point $x = x' + l_* e_i$ on the segment $[x', x' + le_i] = (x', x' + e_i, \ldots, x' + le_i)$ satisfying $\frac{1}{8}l^{d+1} < |C_x(l) \cap I|$. By the above observations, this point x is well-defined and not equal to x'. By (5.21), x then also satisfies

$$(5.22) \qquad \frac{l^{d+1}}{8} < |C_x(l) \cap I|$$
$$\overset{(5.21)}{\leq} |C_{x'+(l_*-1)e_i}(l) \cap I| + \frac{cl^{d+1}}{N^\gamma}$$
$$\leq \left(\frac{1}{8} + \frac{c}{N^\gamma}\right) l^{d+1} \leq \frac{l^{d+1}}{7},$$

for $N \geq c(\gamma)$. In addition, one has $\pi_i(x) = \pi_i(x') = w$. This construction thus yields, for any $w \in \pi_i(\partial_{\mathcal{C}_l,i}(\mathcal{A}))$, a point $x \in C(L)^{(-l)}$ (note that $x', x' + le_i \in C(L)^{(-l)}$ and $C(L)^{(-l)}$ is convex), satisfying (5.22) and $\pi_i(x) = w$. We define the set $\mathcal{E}'$ as the set of all such points x. Then by construction, we have $\pi_i(\mathcal{E}') = \pi_i(\partial_{\mathcal{C}_l,i}(\mathcal{A}))$, in particular (5.12) holds with $\mathcal{E}'$ in place of $\mathcal{E}$ and $\pi_* = \pi_i$, as does (5.13), by (5.20). For any $x \in \mathcal{E}'$, we apply the isoperimetric inequality (5.11) of Lemma 5.2 with $C_x(l)$ in place of $[0, M]^{d+1}$, $C_x(l) \cap I$ in place of A and $1 - \kappa = \frac{1}{7}$, cf. (5.22). We thus find a $j(x) \in \{1, \ldots, d+1\}$ with

$$(5.23) \qquad |\pi_{j(x)}(\partial_{C_x(l),j(x)}(C_x(l) \cap I))| \geq c|C_x(l) \cap I|^{\frac{d}{d+1}} \overset{(5.22)}{\geq} c'l^d.$$

It follows from the choice of I that $\partial_{C_x(l),j(x)}(C_x(l) \cap I) \subseteq K \cap C_x(l)$, and hence

$$(5.24) \qquad |\pi_{j(x)}(K \cap C_x(l))| \geq cl^d.$$

We now let π_{**} be the $\pi_{j(x)}$ occurring most in (5.23), where x varies over $\mathcal{E}'$, and define $\mathcal{E} \subseteq \mathcal{E}'$, as the subset of those x in $\mathcal{E}'$ for which $\pi_{j(x)} = \pi_{**}$. With this choice, (5.14) holds by (5.24). Moreover, since (5.12) and (5.13) both hold for $\mathcal{E}'$ and since $|\mathcal{E}| \geq \frac{1}{d+1}|\mathcal{E}'|$, the same identities hold for $\mathcal{E}$ as well (with a different constant). Hence, the proof of Lemma 5.3 is complete under (5.19).

52   1. Disconnection of a discrete cylinder by a biased random walk

On the other hand, let us now assume (5.19) does not hold. That is, we suppose that

(5.25) $$|\mathcal{A}| > c_4|\mathcal{C}_l|.$$

We then claim that, for $N \geq c(\gamma, \gamma')$,

(5.26) $$\left|\{x \in \mathcal{A} : |C_x(l) \cap I| > c_4 l^{d+1}\}\right| \leq c_4|\mathcal{A}|.$$

Indeed, we would otherwise have

$$|I| \geq \left|\{x \in \mathcal{A} : |C_x(l) \cap I| > c_4 l^{d+1}\}\right| c_4 l^{d+1} \stackrel{\text{(if (5.26) false)}}{>} c_4^2|\mathcal{A}|l^{d+1}$$
$$\stackrel{(5.25)}{>} c_4^3 l^{d+1}|\mathcal{C}_l| \stackrel{(5.19)}{>} \frac{4}{5}l^{d+1}|\mathcal{C}_l| = \frac{4}{5}\left(\frac{l^{d+1}|\mathcal{C}_l|}{L^{d+1}}\right)L^{d+1},$$

contradicting the choice of I for $N \geq c(\gamma, \gamma')$, because $\frac{l^{d+1}|\mathcal{C}_l|}{L^{d+1}}$ only depends on N, γ, γ' and tends to 1 as $N \to \infty$. It follows that for $N \geq c(\gamma, \gamma')$,

(5.27) $$c_4|\mathcal{C}_l| \stackrel{(5.25)}{\leq} |\mathcal{A}| \stackrel{(5.26)}{\leq} \frac{1}{1-c_4}\left|\{x \in \mathcal{A} : |C_x(l) \cap I| \leq c_4 l^{d+1}\}\right|$$
$$\stackrel{(5.16)}{=} \frac{1}{1-c_4}\left|\left\{x \in \mathcal{C}_l : \frac{1}{8}l^{d+1} < |C_x(l) \cap I| \leq c_4 l^{d+1}\right\}\right|.$$

Defining $\mathcal{E}' = \{x \in \mathcal{C}_l : \frac{1}{8}l^{d+1} < |C_x(l) \cap I| \leq c_4 l^{d+1}\}$, we apply the isoperimetric inequality (5.11) of Lemma 5.2 with $C_x(l)$ in place of $[0, M]^{d+1}$ and $C_x(l) \cap I$ in place of A for every $x \in \mathcal{E}'$ and thus obtain a projection $\pi_{j(x)}$ satisfying (5.24), as in the previous case. We then define $\mathcal{E} \subseteq \mathcal{E}'$ as the subset containing only those $x \in \mathcal{E}'$ for which $\pi_{j(x)}$ in (5.24) is equal to the most frequently occurring π_{**}. As a consequence, (5.14) holds. Moreover, (5.12) is clear by definition of $\mathcal{E}$. And finally, we have by (5.27), $|\mathcal{E}| \geq \frac{1}{d+1}|\mathcal{E}'| \geq c|\mathcal{C}_l| \geq c'\left(\frac{L}{l}\right)^{d+1}$, which yields (5.13) by (5.9). This completes the proof of Lemma 5.3. □

The last geometric lemma in this section is essentially a modification of Lemma 5.3. It provides a similar result for $B(\alpha)$, $0 < d\alpha < 1$ instead of $C(L)$. The idea of the proof, illustrated in Figure 4, is to "pile up" approximately $N^{1-d\alpha}$ copies of $B(\alpha)$ into the $\mathbb{Z}$-direction of E and to then apply the same arguments with the isoperimetric inequality (5.11) as in the proof of Lemma 5.3 to the resulting set intersected with $B_\infty(x, [N/4])$.

LEMMA 5.4. $(d \geq 1, 0 < \gamma < d\alpha < 1, l = [N^\gamma], N \geq 1)$
For all $N \geq c(\alpha, \gamma)$, whenever $K \subseteq B(\alpha)$ (cf. (1.7)) $\frac{1}{3}$-disconnects $B(\alpha)$, there exists a set $\mathcal{E} \subseteq B(\alpha)^{(-l)}$ and projections π_* and $\pi_{**} \in$

5. More geometric lemmas

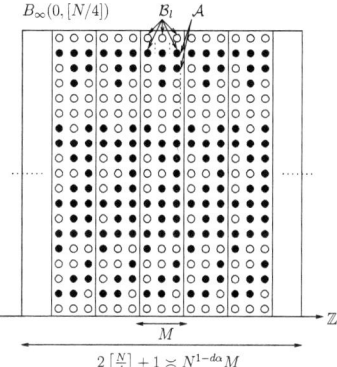

Figure 4: An illustration of the set $\mathcal{A}'$ of copies of $\mathcal{A} \subseteq \mathcal{B}_l$ piled up in the (horizontal) $\mathbb{Z}$-direction (cf. (5.35)), used in the proof of Lemma 5.4. The circles are the points on the subgrid $\mathcal{H}_l$ in (5.31), and the filled circles are the points contained in the set $\mathcal{A}'$. Each copy of $\mathcal{B}_l$ has thickness M, defined in (5.34), so that the larger box $B_\infty(0, [N/4])$ contains roughly $N^{1-d\alpha}$ copies of $\mathcal{B}_l$.

$\{\pi_1, \ldots, \pi_{d+1}\}$ such that

(5.28) $\pi_*(\mathcal{E}) \subseteq \pi_* \left(B(\alpha) \cap \pi_E \left([-[N/4], [N/4]]^{d+1} \cap l\mathbb{Z}^{d+1} \right) \right),$

(5.29) $|\pi_*(\mathcal{E})| \geq c' \left(\frac{N}{l} \right)^d N^{d\alpha - 1},$ and

(5.30) for all $x \in \mathcal{E} : |\pi_{**}(K \cap C_x(l))| \geq c'' l^d.$

PROOF. The proof is very similar to the one of Lemma 5.3. We choose a set $I \subseteq B(\alpha)$ such that $\frac{1}{3}|B(\alpha)| \leq |I| \leq \frac{2}{3}|B(\alpha)|$ and $\partial_{B(\alpha)}(I) \subseteq K$. We then introduce the subgrids of side-length l of $[-[N/4], [N/4]]^d \times \mathbb{Z}$ and of $B(\alpha)^{(-l)}$ as (cf. (2.3))

(5.31) $\mathcal{H}_l = \pi_E \left(\left([-[N/4], [N/4]]^d \times \mathbb{Z} \right) \cap l\mathbb{Z}^{d+1} \right)$ and
$\mathcal{B}_l = B(\alpha)^{(-l)} \cap \mathcal{H}_l,$

and set

$$\mathcal{A} = \left\{ x \in \mathcal{B}_l : |C_x(l) \cap I| > \frac{1}{6} l^{d+1} \right\}.$$

Since the disjoint union $\bigcup_{x \in \mathcal{B}_l} C_x(l)$ contains all but at most $cN^d l$ points of $B(\alpha)$, we then have

$$\frac{1}{3}|B(\alpha)| \leq |I| \leq \frac{1}{6} l^{d+1} |\mathcal{B}_l \setminus \mathcal{A}| + l^{d+1}|\mathcal{A}| + cN^d l$$

$$\leq \frac{1}{6}|B(\alpha)| + l^{d+1}|\mathcal{A}| + cN^d l,$$

hence

$$\left(\frac{1}{6} - cN^{\gamma - d\alpha}\right) \frac{|B(\alpha)|}{l^{d+1}} \leq |\mathcal{A}|,$$

and thus for $N \geq c(\alpha, \gamma)$,

(5.32) $$c|\mathcal{B}_l| \leq \frac{c|B(\alpha)|}{l^{d+1}} \leq |\mathcal{A}|.$$

Suppose now that in addition

(5.33) $$|\mathcal{A}| \leq c_5 |\mathcal{B}_l| \quad \text{with } c_5^3 = \frac{1}{2}\left(1 + \frac{3}{4}\right).$$

Then we define the set $\mathcal{A}' \subseteq \mathcal{H}_l$ by "piling up" adjoining copies of the set $\mathcal{B}_l \supseteq \mathcal{A}$ into the $\mathbb{Z}$-direction. That is, we introduce the "thickness" M of $\mathcal{B}_l$,

(5.34) $$M = \sup_{(u,v),(u',v') \in \mathcal{B}_l} |v - v'| = 2 \left[\frac{\left[\frac{1}{4} N^{d\alpha}\right] - l}{l}\right] l,$$

and define

(5.35) $$\mathcal{A}' = \bigcup_{n \in \mathbb{Z}} (n(M+l)e_{d+1} + \mathcal{A}) \subseteq \bigcup_{n \in \mathbb{Z}} (n(M+l)e_{d+1} + \mathcal{B}_l) = \mathcal{H}_l,$$

cf. (5.31), Figure 4. Observe that $B_\infty(0, [N/4]) \cap \mathcal{A}'$ contains no less than $cN^{1-d\alpha}$ and no more than $c'N^{1-d\alpha}$ copies of $\mathcal{A}$. With (5.32) and (5.33) it follows that for $N \geq c(\alpha, \gamma)$,

$$c'\left(\frac{N}{l}\right)^{d+1} \leq |B_\infty(0, [N/4]) \cap \mathcal{A}'| \leq (1 - c')\left(\frac{N}{l}\right)^{d+1}.$$

For $N \geq c(\gamma')$, an application of the isoperimetric inequality (5.11) of Lemma 5.2 on the subgrid $\mathcal{H}_l$ defined in (5.31), with $B_\infty(0, [N/4]) \cap \mathcal{H}_l$ in place of $[0, M]^{d+1}$ and $B_\infty(0, [N/4]) \cap \mathcal{A}'$ in place of A, hence yields an $i \in \{1, \ldots, d+1\}$ such that

(5.36) $$|\pi_i(\partial_{\mathcal{H}_l, i}(\mathcal{A}'))| \geq c\left(\frac{N}{l}\right)^d.$$

If $i \neq d+1$, then the set on the left-hand side of (5.36) is contained in the at most $cN^{1-d\alpha}$ translated copies of the set $\pi_i(\partial_{\mathcal{B}_l, i}(\mathcal{A}))$ intersecting

6. The large deviation estimate

$B_\infty(0, [N/4])$ (see (5.35) and Figure 4). We then deduce from (5.36) that

(5.37) $$|\pi_i(\partial_{\mathcal{B}_l, i}(\mathcal{A}))| \geq c \left(\frac{N}{l}\right)^d N^{d\alpha - 1}.$$

If $i = d+1$, in (5.36), then we claim that

(5.38) $$\pi_{d+1}(\partial_{\mathcal{B}_l, d+1}(\mathcal{A})) \supseteq \pi_{d+1}(\partial_{\mathcal{H}_l, d+1}(\mathcal{A}')).$$

Indeed, suppose some $u \in \mathbb{T}_N^d$ does not belong to the left-hand side. Then the fiber $\{x \in \mathcal{B}_l : \pi_{d+1}(x) = u\}$ must either be disjoint from $\mathcal{A}$ or be a subset of $\mathcal{A}$. Our construction of $\mathcal{A}'$ in (5.35) implies that the set
$$\{x \in \mathcal{H}_l : \pi_{d+1}(x) = u\}$$
is then either disjoint from $\mathcal{A}'$ or a subset of $\mathcal{A}'$, as in the first and second horizontal lines of Figure 4 (note that the translated copies of $\mathcal{B}_l$ in (5.35) adjoin each other on the subgrid $\mathcal{H}_l$). But this precisely means that u is not included in the right-hand side of (5.38). In particular, by (5.36) and (5.38), (5.37) holds also with $i = d + 1$ (even without the $N^{d\alpha - 1}$ on the right-hand side). Using (5.37), we can perform the same construction as in the proof of Lemma 5.3 below (5.20) in order to obtain the desired set $\mathcal{E}$.

If, on the other hand, (5.33) does not hold, i.e. if
$$|\mathcal{A}| > c_5 |\mathcal{B}_l|,$$
then the existence of the required set $\mathcal{E}$ follows from the argument below (5.25), where (5.29) can be deduced from $|\mathcal{E}| \geq c|\mathcal{B}_l| \geq c'(N/l)^{d+1} N^{d\alpha - 1}$ by applying the estimate (5.9) to $[cN^{1-d\alpha}]$ copies of $\mathcal{E}$ piled-up in a box. □

6. The large deviation estimate

Our task in this last section is to derive the following form of the large deviation estimate (1.10):

THEOREM 6.1. ($d \geq 3$) The estimate (1.10) holds with (cf. Figure 2, p. 28)

(6.1)
$$f(\alpha, \beta) = \begin{cases} d - 1 - \frac{d\alpha}{d-1} & \text{on } (0, 1/d) \times \left(0, d - 1 - \frac{d\alpha}{d-1}\right), \\ d - 1 - \frac{1}{d-1} & \text{on } [1/d, \infty) \times \left(0, d - 1 - \frac{1}{d-1}\right), \\ ((d-1)^2 - 1) \times & \\ (d - 1 - \beta) & \text{on } [1/d, \infty) \times \left[d - 1 - \frac{1}{d-1}, d - 1\right), \\ 0 & \text{otherwise.} \end{cases}$$

Before we begin with the proof of Theorem 6.1, we examine its implications. With the function f in (6.1), the lower bound exponents $d(1 - \alpha - \varphi(\alpha))$ (in (1.4)) and ζ (in (1.13)) are related via (1.14), as will be checked in Corollary 6.3. We therefore have to justify the expression $\vee d(1-2\alpha)\mathbf{1}_{\{\alpha < \frac{1}{d}\}}$ on the right-hand side of (1.14). This is the aim of the next proposition.

PROPOSITION 6.2. ($d \geq 2$, $0 < \alpha < \frac{1}{d}$) For some constant $c_6 > 0$,

(6.2) $\quad P_0^{N-d\alpha}\left[\exp\{c_6 N^{d(1-2\alpha)}\} \leq T_N\right] \stackrel{N \to \infty}{\longrightarrow} 1$.

PROOF. The idea is that, by our previous geometric estimates, any trajectory disconnecting E must contain at least cN^d points in a box of the form $x + B_\infty(0, [N/4])$, $x \in E$. Hence, there must be two visited points within distance N from each other, such that the random walk X spends $[cN^d]$ time units between the visits to the two points. The probability of this event can be bounded from above by standard large deviation estimates.

In detail: Lemma 4.1, applied with $B(\alpha) = B_\infty(0, [N/4])$ (i.e. with $\alpha \geq \frac{1}{d}$), shows that, for $t \geq 0$, $N \geq c$, the event

$$\{X([0,[t]]) \text{ disconnects } E\}$$

is contained in the event

(6.3) $\quad \bigcup_{\substack{x \in E \\ |x_{d+1}| \leq [t]+N}} \left\{X([0,[t]]) \tfrac{1}{3}\text{-disconnects } x + B_\infty(0, [N/4])\right\}$.

We now choose a set $I \subseteq x + B_\infty(0, [N/4])$ corresponding to $\frac{1}{3}$-disconnection of $x + B_\infty(0, [N/4])$ by $X([0,[t]])$ (cf. (1.6)). By the isoperimetric inequality (5.11) of Lemma 5.2, applied with $x + B_\infty(0, [N/4])$ in place of $[0, M]^{d+1}$ and I in place of A, the event (6.3) is contained in $\bigcup_{\substack{x \in E \\ |x_{d+1}| \leq [t]+N}} A_x([t])$, where, for some constant $c_7 > 0$,

$$A_x([t]) = \{|X([0,[t]]) \cap (x + B_\infty(0, [N/4]))| \geq c_7 N^d\}.$$

We therefore have

(6.4) $\quad P_0^{N-d\alpha}[T_N \leq t] \leq P_0^{N-d\alpha}\left[\bigcup_{\substack{x \in E \\ |x_{d+1}| \leq [t]+N}} A_x([t])\right]$

$$\leq cN^d(t+N) \sup_{x \in E} P_0^{N-d\alpha}[A_x([t])].$$

By the strong Markov property applied at $H^X_{x+B_\infty(0,[N/4])}$, the entrance time of $x + B_\infty(0, [N/4])$, and using translation invariance of X, we obtain

$$\sup_{x \in E} P_0^{N-d\alpha}[A_x([t])] \leq \sup_{x \in E} P_0^{N-d\alpha}\left[\theta^{-1}_{H^X_{(x+B_\infty(0,[N/4]))}} A_x([t])\right]$$

(Markov, transl. inv.)

$$\leq \sup_{x: A_x([t]) \ni 0} P_0^{N-d\alpha}[A_x([t])]$$

$$\leq P_0^{N-d\alpha}\left[\text{for some } n \geq c_7 N^d : \pi_{\mathbb{Z}}(X_n) \leq N\right].$$

6. The large deviation estimate

Inserting this last inequality into (6.4) and using that $N \leq \frac{N^{-d\alpha}n}{2(d+1)}$ for $n \geq c_7 N^d$, $N \geq c(\alpha)$ (because $d - d\alpha > d - 1 \geq 1$), we deduce that, for $N \geq c(\alpha)$,

$$
\begin{aligned}
(6.5) \quad P_0^{N^{-d\alpha}} [T_N \leq t] \\
&\leq cN^d (t+N) \, P_0^{N^{-d\alpha}} \left[\text{for some } n \geq c_7 N^d : \pi_{\mathbb{Z}}(X_n) \leq N \right] \\
&\leq cN^d (t+N) \sum_{n \geq c_7 N^d} P_0^{N^{-d\alpha}} \left[\pi_{\mathbb{Z}}(X_n) \leq \frac{N^{-d\alpha}n}{2(d+1)} \right] \\
&= cN^d (t+N) \sum_{n \geq c_7 N^d} P_0^{N^{-d\alpha}} \left[\pi_{\mathbb{Z}}(X_n) - \frac{N^{-d\alpha}n}{d+1} < -\frac{N^{-d\alpha}n}{2(d+1)} \right].
\end{aligned}
$$

Since $\left(\pi_{\mathbb{Z}}(X_n) - \frac{N^{-d\alpha}n}{d+1} \right)_{n \geq 0}$ is a $P_0^{N^{-d\alpha}}$-martingale with steps bounded by c, Azuma's inequality (cf. [4], p. 85) implies that

$$ P_0^{N^{-d\alpha}} \left[\pi_{\mathbb{Z}}(X_n) - \frac{N^{-d\alpha}n}{d+1} < -\frac{N^{-d\alpha}n}{2(d+1)} \right] \leq e^{-cN^{-2d\alpha}n}. $$

Applying this estimate to (6.5) with $t_N = \exp\{c_6 N^{d-2d\alpha}\}$ we see that for $N \geq c(\alpha)$,

$$ P_0^{N^{-d\alpha}} \left[T_N \leq \exp\{c_6 N^{d-2d\alpha}\} \right] \leq cN^{d+2d\alpha} \exp\left\{ c_6 N^{d-2d\alpha} - c' N^{d-2d\alpha} \right\}. $$

Choosing the constant $c_6 > 0$ sufficiently small, this yields (6.2) (recall that $d\alpha < 1 \leq \frac{d}{2}$). $\square$

We can now check that Theorem 6.1 does have the desired implications on the lower bounds on T_N.

COROLLARY 6.3. ($d \geq 3$, $\alpha > 0$, $\epsilon > 0$) With φ defined in (1.5), one has

$$(6.6) \quad \text{for } \alpha > 1, \quad P_0^{N^{-d\alpha}} \left[N^{2d-\epsilon} \leq T_N \right] \xrightarrow{N \to \infty} 1,$$

$$(6.7) \quad \text{for } \alpha < 1, \quad P_0^{N^{-d\alpha}} \left[\exp\{N^{d(1-\alpha-\varphi(\alpha))-\epsilon}\} \leq T_N \right] \xrightarrow{N \to \infty} 1.$$

PROOF. Since the function f of (6.1) satisfies $f(\alpha, \beta) > 0$ for $(\alpha, \beta) \in (1, \infty) \times (0, d-1)$, (6.6) follows immediately from Theorem 6.1 and (1.11).

By (1.12) and (6.2), (6.7) holds with φ defined for $\alpha \in (0, 1)$ by (1.14). Let us check that the expression for φ in (1.14) agrees with (1.5). We first treat the case $\alpha \in [\frac{1}{d}, 1)$, for which $f(\alpha, .)$ is illustrated on the right-hand side of Figure 2, below Theorem 1.2. We have $d\alpha \geq 1$, $f(\alpha, \beta) = 0$ for $\beta \geq d-1$ and the maximum of g_α (cf. (1.13)) on $(0, d-1)$ is attained at (see Figure 2)

$$ \bar{\beta} = d - 1 - \frac{d - d\alpha}{(d-1)^2} \in \left[d - 1 - \frac{1}{d-1}, d - 1 \right) \cap (d\alpha - 1, d - 1). $$

Hence, for $\alpha \in [\frac{1}{d}, 1)$, cf. (1.13),

$$\zeta = \sup_{\beta > 0} g_\alpha(\beta) = g_\alpha(\bar{\beta}) = d\left(1 - \alpha - \frac{1-\alpha}{(d-1)^2}\right),$$

and therefore $\varphi(\alpha) = \frac{1-\alpha}{(d-1)^2}$, as required.

Turning to the case $\alpha \in (0, \frac{1}{d})$, we refer to the left-hand side of Figure 2 for an illustration of f. We now have $d\alpha - 1 < 0$, $f(\alpha, \beta) = 0$ for $\beta > d - 1 - \frac{d\alpha}{d-1}$ and hence

$$\zeta = \sup_{\beta > 0} g_\alpha(\beta) = \sup_{\beta \in \left(0, d-1-\frac{d\alpha}{d-1}\right)} \left(\beta \wedge d - 1 - \frac{d\alpha}{d-1}\right)$$

$$= d\left(1 - \frac{1}{d} - \frac{\alpha}{d-1}\right).$$

Therefore (cf. (1.14)), for $\alpha \in (0, \frac{1}{d})$,

(6.8) $\qquad \varphi(\alpha) = 1 - \alpha - \left(\left(1 - \frac{1}{d} - \frac{\alpha}{d-1}\right) \vee (1 - 2\alpha)\right).$

This expression is immediately seen to coincide with (1.5) for $\alpha \in (0, \frac{1}{d})$ near 0 and $\frac{1}{d}$, and α_* is precisely the value for which $1 - \frac{1}{d} - \frac{\alpha_*}{d-1} = 1 - 2\alpha_*$, so that (6.8), and hence (1.14), agrees with (1.5). □

Thanks to Corollary 6.3, the lower bounds on T_N of Theorem 1.1 will be established once we show Theorem 6.1. Let us give a rough outline of the strategy of the proof. In the previous section, we have shown that if $K = X\left([0, \mathcal{D}_{[N^\beta]}]\right) \frac{1}{3}$-disconnects $B(\alpha)$, then there must be a wealth of subcubes of $B(\alpha)$ such that $X\left([0, \mathcal{D}_{[N^\beta]}]\right)$ contains a surface of points in every subcube (see Lemmas 5.3 and 5.4 for the precise statements and Figure 3 for an illustration). The crucial upper bound on the probability of an event of this form is obtained in Lemma 6.5, using Khaśminskii's Lemma to obtain an exponential tail estimate on the number of points visited by X during a suitably defined excursion. This upper bound is then applied in order to find the needed large deviation estimate of the form (1.10). We begin by collecting the required estimates involving the Green function (cf. (2.20)).

LEMMA 6.4. $(d \geq 2, N, a \geq 1, 100 \leq a \leq 4N, A \subseteq B \subseteq S_a)$

(6.9) $\qquad P_x^0\left[H_A^X < H_{B^c}^X\right] \leq \dfrac{\sum_{y \in A} g^B(x, y)}{\inf_{y \in A} \sum_{y' \in A} g^B(y, y')},$ for $x \in B$.

For any $x, x' \in S_a$, one has

(6.10) $\qquad g^{S_a}(x, x') \leq c\left(1 \vee |x - x'|_\infty\right)^{1-d} \exp\left\{-c'\dfrac{|x - x'|_\infty}{a}\right\}.$

6. The large deviation estimate

If $diam(A) \le \frac{a}{100}$ (cf. (2.4)) and $A \subseteq B^{-(\frac{a}{10})}$ (cf. (2.3)), then, for $x, x' \in A$,

(6.11) $$c|x - x'|_\infty^{1-d} \le g^B(x, x').$$

PROOF. The estimate (6.9) follows from an application of the strong Markov property at H_A^X. The estimate (6.10) follows from the bound on the Green function of the simple random walk on $\mathbb{Z}^{d+1}$ killed when exiting the slab $\mathbb{Z}^d \times [-[a], [a]]$ in (2.13) of Sznitman [34]. For (6.11), we note that, by assumption, $B_\infty\left(x, \frac{a}{10}\right) \subseteq B$. In particular, it follows from translation invariance that

(6.12) $$g^B(x, x') \ge g^{B_\infty\left(0, \frac{a}{10}\right)}(0, x - x').$$

By assumption $\frac{a}{10} \le \frac{2N}{5}$, so the right-hand side of (6.12) can be identified with the corresponding Green function for the simple random walk on $\mathbb{Z}^{d+1}$, and (6.11) follows from Lawler [19], p. 35, Proposition 1.5.9. □

We now introduce, for sets $U, \tilde{U} \subseteq E$, the the times $(\tilde{R}_n)_{n \ge 1}$ and $(\tilde{D}_n)_{n \ge 1}$ as the times of return to U and departure from $\tilde{U}$ (cf. (2.19)) and denote with π_* and π_{**} elements of the set of projections

$$\{\pi_1, \ldots \pi_{d+1}\}.$$

The next lemma then provides a control on an event of the form (cf. (2.3), (2.8))

(6.13) $$A_{U,\tilde{U},l,M_1,M_2} = \bigcup_{\pi_*, \pi_{**}} \bigcup_{\substack{\mathcal{E} \subseteq U^{(-l)}: \\ |y-y'|_\infty \ge l \text{ for } y,y' \in \mathcal{E}, \\ |\pi_*(\mathcal{E})| \ge M_1}} \bigcap_{y \in \mathcal{E}} \left\{\left|\pi_{**}\left(X([0, \tilde{D}_{M_2}]) \cap C_y(l)\right)\right| \ge cl^d\right\}.$$

Our method does not produce a useful upper bound when $d = 2$ (note that when $d = 2$, the right-hand side of (6.14) is greater than 1 for $N \ge c$). Although it is possible to obtain a bound for $d = 2$ tending to 0 as $N \to \infty$, using estimates on the Green function in dimension 2, it does not seem to be possible to obtain an exponential decay in N with this approach. Thus, the upper bound we have for $d = 2$ brings little information on the large deviation problem (1.10).

LEMMA 6.5. ($d \ge 3$, $N, l, a, M_1, M_2 \ge 1$, $100 \le a \le 4N$, $1 \le l \le \frac{a}{100}$) Let $U, \tilde{U} \subseteq E$ be sets such that $U \subseteq U^{(\frac{a}{10})} \subseteq \tilde{U} \subseteq x_* + S_a$ (cf. (2.2), (1.9)). Then one has the estimate

(6.14) $$\sup_{x \in E} P_x^0\left[A_{U,\tilde{U},l,M_1,M_2}\right]$$
$$\le \exp\left\{c'M_2 + c'M_1 \log N - c''M_1 a^{-1} l^{d-1}\right\}$$

(on the event defined in (6.13)).

PROOF. In order to abbreviate the notation, we denote the event in (6.13) by A during the proof. Furthermore, by replacing $\mathcal{E}$ with a subset, we may assume that

(6.15) $$|\pi_*(\mathcal{E})| = |\mathcal{E}| = M_1.$$

Also, translation invariance allows us to set $x_* = 0$.

The first step is to note that the number of possible choices of the set $\mathcal{E}$ in the definition of A is not larger than

$$|U|^{|\mathcal{E}|} \overset{(6.15)}{\leq} (cN)^{(d+1)M_1} \leq \exp\{cM_1 \log N\}.$$

Next, we note that visits made by the random walk X to $C_y(l)$, $y \in \mathcal{E}$, can only occur during the time intervals $[\tilde{R}_n, \tilde{D}_n]$, $n \geq 1$ (because $\mathcal{E} \subseteq U^{(-l)}$). From these observations, we deduce that $\sup_{x \in E} P_x^0[A]$ is bounded by

$$ce^{cM_1 \log N} \sup_{x, \mathcal{E}, \pi_*, \pi_{**}} P_x^0\left[\sum_{n=1}^{M_2} \sum_{y \in \mathcal{E}} \left|\pi_{**}\left(X([\tilde{R}_n, \tilde{D}_n]) \cap C_y(l)\right)\right| \geq cM_1 l^d\right],$$

where the supremum is taken over all $x \in E$, and all possible sets $\mathcal{E}$ and projections π_*, π_{**} entering the definition of the event A. By the exponential Chebychev inequality and the strong Markov property applied inductively at $\tilde{R}_{M_2}, \tilde{R}_{M_2-1}, \ldots, \tilde{R}_1$, it follows from the last estimate that, for any $r \geq 1$, $\sup_{x \in E} P_x^0[A]$ is bounded by

(6.16)
$$ce^{cM_1 \log N - crM_1 l^d} \times$$

$$\sup_{x, \mathcal{E}, \pi_*, \pi_{**}} E_x^0\left[\exp\left\{\sum_{n=1}^{M_2} \sum_{y \in \mathcal{E}} r\left|\pi_{**}\left(X([0, \tilde{D}_1]) \cap C_y(l)\right)\right| \circ \theta_{\tilde{R}_n}\right\}\right]$$

$$\overset{(\text{Markov})}{\leq} ce^{cM_1 \log N - crM_1 l^d} \times$$

$$\sup_{\mathcal{E}, \pi_*, \pi_{**}} \left(\sup_{x \in U} E_x^0\left[\exp\left\{\sum_{y \in \mathcal{E}} r\left|\pi_{**}\left(X([0, \tilde{D}_1]) \cap C_y(l)\right)\right|\right\}\right]\right)^{M_2}.$$

Before deriving an upper bound on this last expectation, we introduce the following notational simplification: for any point $z \in C_y(l)$, we denote its fiber in $C_y(l)$ of points of equal π_{**}-projection by J_z, or in other words, for $z \in C_y(l)$,

$$J_z = \{z' \in C_y(l) : \pi_{**}(z') = \pi_{**}(z)\}.$$

The collection of all fibers in the box $C_y(l)$ is denoted by

(6.17) $$F(y) = \{J_z : z \in C_y(l)\},$$

6. The large deviation estimate

and the collection of all fibers by

(6.18) $$F = \bigcup_{y \in \mathcal{E}} F(y).$$

Using this notation, we have (cf. (2.17))

(6.19) $$\sum_{y \in \mathcal{E}} \left| \pi_{**}\left(X([0, \tilde{D}_1]) \cap C_y(l)\right) \right| = \sum_{J \in F} \mathbf{1}_{\{H_J^X < \tilde{D}_1\}}.$$

By the version of Khaśminskii's Lemma of equation (2.46) of Dembo and Sznitman [12], see also [18], we see that for any $x \in U$ and $r \geq 0$,

(6.20) $$E_x^0\left[\exp\left\{r \sum_{J \in F} \mathbf{1}_{\{H_J^X < \tilde{D}_1\}}\right\}\right]$$
$$\leq \sum_{k \geq 0} r^k \left(\sup_{x \in U} E_x^0\left[\sum_{J \in F} \mathbf{1}_{\{H_J^X < \tilde{D}_1\}}\right]\right)^k.$$

Writing (cf. (6.17), (6.18))

$$\sum_{J \in F} \mathbf{1}_{\{H_J^X < \tilde{D}_1\}} = \sum_{y \in \mathcal{E}} \sum_{J \in F(y)} \mathbf{1}_{\{H_J^X < \tilde{D}_1\}},$$

for any $x \in U$, the strong Markov property applied at $H_{C_y(l)}^X$ yields

(6.21) $$E_x^0\left[\sum_{J \in F} \mathbf{1}_{\{H_J^X < \tilde{D}_1\}}\right]$$
$$= \sum_{y \in \mathcal{E}} E_x^0\left[H_{C_y(l)}^X < \tilde{D}_1, \sum_{J \in F(y)} \left(\mathbf{1}_{\{H_J^X < \tilde{D}_1\}}\right) \circ \theta_{H_{C_y(l)}^X}\right]$$
$$\leq \sum_{y \in \mathcal{E}} P_x^0\left[H_{C_y(l)}^X < \tilde{D}_1\right] \sup_{z \in C_y(l)} E_z^0\left[\sum_{J \in F(y)} \mathbf{1}_{\{H_J^X < \tilde{D}_1\}}\right].$$

To bound the right-hand side of (6.21), we note that, for any $z \in C_y(l)$ and $k \in \{0, \ldots, l-1\}$, at most $c(1 \vee k)^{d-1}$ of the fibers $J \in F(y)$ are at $|.|_\infty$-distance $1 \vee k$ from J_z and thus deduce that, for any $z \in C_y(l)$,

(6.22) $$E_z^0\left[\sum_{J \in F(y)} \mathbf{1}_{\{H_J^X < \tilde{D}_1\}}\right]$$
$$\leq c \sum_{k=0}^{l-1} (1 \vee k)^{d-1} \sup_{z' : |\pi_{**}(z-z')|_\infty = k} P_z^0\left[H_{J_{z'}}^X < \tilde{D}_1\right].$$

For this last probability, we use the estimate (6.9), applied with $A = J_{z'}$, $B = \tilde{U}$ and $x = z$. With the help of (6.10) and the assumption that $\tilde{U} \subseteq S_a$, the numerator of the right-hand side of (6.9) can then be bounded from above by clk^{1-d}, while the denominator is trivially bounded from below by 1. We thus obtain

$$\sup_{z':|\pi_{**}(z-z')|=k} P_z^0\left[H_{J_{z'}}^X < \tilde{D}_1\right] \leq clk^{1-d}.$$

With (6.22), this yields

$$E_z^0\left[\sum_{J\in F(y)} \mathbf{1}_{\{H_J^X < \tilde{D}_1\}}\right] \leq cl^2, \text{ for any } z \in C_y(l).$$

Coming back to (6.21), we obtain, for any $x \in U$,

(6.23) $$E_x^0\left[\sum_{J\in F} \mathbf{1}_{\{H_J^X < \tilde{D}_1\}}\right] \leq cl^2 \sum_{y\in\mathcal{E}} P_x^0\left[H_{C_y(l)}^X < \tilde{D}_1\right].$$

For this last sum, we proceed as before: Note that, by (6.15), the sum can be regarded as a sum over the set $\pi_*(\mathcal{E})$, which is a subset of the d-dimensional lattice $\pi_*(E)$. Since moreover $|y-y'|_\infty \geq l$ for all $y, y' \in \mathcal{E}$, there are at most $c(1\vee k)^{d-1}$ points in $\pi_*(\mathcal{E})$ of $|.|_\infty$-distance between kl and $(k+1)l$ from $\pi_*(x)$. We therefore have, for any $x \in U$,

(6.24) $$\sum_{y\in\mathcal{E}} P_x^0\left[H_{C_y(l)}^X < \tilde{D}_1\right]$$

$$\leq c\sum_{k=0}^\infty (1\vee k)^{d-1} \sup_{y\in\mathcal{E}:|\pi_*(y-x)|\geq kl} P_x^0\left[H_{C_y(l)}^X < \tilde{D}_1\right].$$

In order to bound this last probability, we again use the estimate (6.9), this time with $A = C_y(l)$ and $B = \tilde{U}$. By (6.10), our assumption that $\tilde{U} \subseteq S_a$ then allows us to bound the numerator of the right-hand side of (6.9) from above by $cl^{d+1}(1\vee lk)^{1-d}e^{-c'lk/a}$, while our assumptions $l \leq \frac{a}{100}$ and $C_y(l) \subseteq U \subseteq \tilde{U}^{(-\frac{a}{10})}$ allow us to use (6.11) and find the lower bound of cl^2 on the denominator. We thus have

$$\sup_{y\in\mathcal{E}:|\pi_*(y-x)|\geq lk} P_x^0\left[H_{C_{x'}(l)}^X < \tilde{D}_1\right] \leq c(1\vee k)^{1-d}e^{-c'\frac{lk}{a}}.$$

With (6.24), this yields

$$\sum_{y\in\mathcal{E}} P_x^0\left[H_{C_y(l)}^X < \tilde{D}_1\right] \leq c\frac{a}{l}, \text{ for any } x \in U,$$

which we insert into (6.23) to obtain

$$\sup_{x\in U} E_x^0\left[\sum_{J\in F} \mathbf{1}_{\{H_J^X < \tilde{D}_1\}}\right] \leq c_8 al.$$

Choosing $r = \frac{1}{2c_8 al}$ in (6.20), we infer that

$$E_x^0\left[\exp\left\{\frac{1}{2c_8 al}\sum_{J\in F}\mathbf{1}_{\{H_J^X<\tilde{D}_1\}}\right\}\right] \leq 2, \text{ for any } x \in U.$$

6. The large deviation estimate

Coming back to (6.16) with r as above and remembering (6.19), we deduce (6.14) and thus complete the proof of Lemma 6.5. □

The remaining part of the proof of Theorem 6.1 is essentially an application of Lemma 6.5 together with the geometric Lemmas 5.1-5.4 showing that the event on the left-hand side of (1.10) is contained in a union of events of the form (6.13). For $\alpha < \frac{1}{d}$, all that remains to be done is to combine Lemma 5.5 with Lemma 6.5. For $\alpha \geq \frac{1}{d}$, i.e. for $B(\alpha) = B_\infty(0, [N/4])$, we additionally use an upper bound on the probability that the random walk X makes a certain number of excursions between $C_{x_*}(L)$ and $\left(C_{x_*}(L)^{(L)}\right)^c$ (cf. (2.8)) until time $\mathcal{D}_{[N^\beta]}$ for $x_* \in B_\infty(0, [N/4])$ and L as in (5.1) (cf. Lemma 6.6) before we apply the geometric Lemmas 5.1 and 5.3 and the estimate (6.14) with $U = C_{x_*}(L)$.

PROOF OF THEOREM 6.1 - CASE $\alpha < \frac{1}{d}$. In this case, we have to show (1.10) with f illustrated on the left-hand side of Figure 2 (below Theorem 1.2) and $[N^{d\alpha \wedge 1}] = [N^{d\alpha}]$. Lemma 5.4 implies that, for l as in (5.1) and the event $A_{.,.,.,.,.}$ defined in (6.13),

$$(6.25) \quad \{U_{B(\alpha)} \leq \mathcal{D}_{[N^\beta]}\} \subseteq A_{S_{2[N^{d\alpha}]}, S_{4[N^{d\alpha}]}, l, cN^{d-1+d\alpha}l^{-d}, [N^\beta]} \stackrel{(\text{def.})}{=} A'_N.$$

Lemma 6.5, applied with $a = 4[N^{d\alpha}]$ and $x_* = 0$, yields

$$(6.26) \quad \sup_{x \in S_{2N^{d\alpha}}} P^0_x[A'_N]$$
$$\leq \exp\left\{cN^\beta + cN^{d-1+d\alpha-d\gamma}\log N - c'N^{d-1-\gamma}\right\}.$$

In view of (5.1), we have $0 < \gamma < d\alpha$, and provided $d - 1 + d\alpha - d\gamma < d - 1 - \gamma$ and $\beta < d - 1 - \gamma$, i.e. if

$$(6.27) \quad \frac{d\alpha}{d-1} < \gamma < d\alpha, \quad \beta < d - 1 - \gamma,$$

then (6.25) and (6.26) together show that

$$(6.28) \quad \sup_{x \in S_{2[N^{d\alpha}]}} P^0_x\left[U_{B(\alpha)} \leq \mathcal{D}_{[N^\beta]}\right] \leq \exp\left\{-cN^{d-1-\gamma}\right\}.$$

For $\beta \in \left(0, d - 1 - \frac{d\alpha}{d-1}\right)$, $d \geq 3$, the constraints (6.27) are satisfied by $\gamma_0 = \frac{d\alpha}{d-1} + \epsilon_0(d, \beta)$ for $\epsilon_0(d, \beta) > 0$ sufficiently small. Moreover, $d - 1 - \gamma_0 = d - 1 - \frac{d\alpha}{d-1} - \epsilon_0(d, \beta) \stackrel{(6.1)}{=} f(\alpha, \beta) - \epsilon_0(d, \beta)$. Since we can make $\epsilon_0(d, \beta) > 0$ arbitrarily small, (6.28) thus shows (1.10) for the case $\alpha \in \left(0, \frac{1}{d}\right)$. This completes the proof of Theorem 6.1 in the case $\alpha < \frac{1}{d}$. □

PROOF OF THEOREM 6.1 - CASE $\alpha \geq \frac{1}{d}$. Recall that in this case we have to find an estimate of the form (1.10) with the function f illustrated on the right-hand side of Figure 2 (below Theorem 1.2) and

64   1. Disconnection of a discrete cylinder by a biased random walk

with $B(\alpha) = B_\infty(0, [N/4])$ (cf. (1.7)). In order to apply Lemma 6.5 with $U = C_{x_*}(L) \subseteq B(\alpha)$, L as in (5.1), we consider

(6.29) $\quad \tilde{R}_n^{x_*}$, $\tilde{D}_n^{x_*}$, $n \geq 1$, the successive returns to $C_{x_*}(L)$ and departures from $C_{x_*}(L)^{(L)}$ (cf. (2.2), (2.19)).

The following lemma, in the spirit of Dembo and Sznitman [12], provides an estimate on the number of excursions between $C_{x_*}(L)$ and $(C_{x_*}(L)^{(L)})^c$ occurring during the $[N^\beta]$ excursions under consideration in (1.10).

LEMMA 6.6. ($d \geq 2$, $\alpha \geq \frac{1}{d}$, $\beta > 0$, $\gamma' \in (0,1)$, $L = [N^{\gamma'}]$, $m, N \geq 1$) For $x \in S_{2N}$, $x_* \in B_\infty(0, [N/4])$ and $\tilde{R}_m^{x_*}$ defined in (6.29),

(6.30) $\quad P_x^0 \left[\tilde{R}_m^{x_*} \leq \mathcal{D}_{[N^\beta]} \right] \leq c \exp \left\{ c N^{1-d} L^{d-1} N^\beta - c' m \right\}.$

PROOF. We follow the proof of Lemma 2.3 by Dembo and Sznitman [12]. Since $C_{x_*}(L) \subseteq S_{2N}$, visits made by X to $C_{x_*}(L)$ can only occur during the time intervals $[\mathcal{R}_i, \mathcal{D}_i]$, $i \geq 1$, cf. above (1.10). Let us denote the number of excursions between $C_{x_*}(L)$ and $(C_{x_*}(L)^{(L)})^c$ performed by X during $[\mathcal{R}_i, \mathcal{D}_i]$ by $\mathcal{N}_i$, i.e.

$$\mathcal{N}_i = \left| \left\{ n \geq 1 : \mathcal{R}_i \leq \tilde{R}_n^{x_*} \leq \mathcal{D}_i \right\} \right|, \quad i \geq 1.$$

Note that one then has $\mathcal{N}_i = \mathcal{N}_1 \circ \theta_{\mathcal{R}_i}$, $i \geq 1$. For any $\lambda > 0$, $x \in S_{2N}$, $x_* \in B_\infty(0, [N/4])$, we apply the strong Markov property at $\mathcal{R}_2$ and deduce that

$$P_x^0 \left[\tilde{R}_m^{x_*} \leq \mathcal{D}_{[N^\beta]} \right]$$
$$\leq P_x^0 \left[\left\{ \mathcal{N}_1 \geq \left[\tfrac{m}{2}\right] \right\} \cup \theta_{\mathcal{R}_2}^{-1} \left\{ \mathcal{N}_1 + \ldots + \mathcal{N}_{[N^\beta]-1} \geq \left[\tfrac{m}{2}\right] \right\} \right]$$
$$\leq P_x^0 \left[\mathcal{N}_1 \geq \left[\tfrac{m}{2}\right] \right] + \sup_{x \in S_{2N} : |x_{d+1}|=2N} P_x^0 \left[\sum_{i=1}^{[N^\beta]-1} \mathcal{N}_i \geq \left[\tfrac{m}{2}\right] \right].$$

With the strong Markov property applied inductively at

$$\mathcal{R}_{[N^\beta]-1}, \mathcal{R}_{[N^\beta]-2}, \ldots, \mathcal{R}_1$$

to the second term on the right-hand side, one infers that

(6.31) $\quad P_x^0 \left[\tilde{R}_m^{x_*} \leq \mathcal{D}_{[N^\beta]} \right]$
$$\leq e^{-\lambda [\frac{m}{2}]} \left(E_x^0 \left[e^{\lambda \mathcal{N}_1} \right] + \sup_{x \in S_{2N} : |x_{d+1}|=2N} E_x^0 \left[e^{\lambda \mathcal{N}_1} \right]^{([N^\beta]-1)} \right).$$

For any $x \in S_{2N}$,

(6.32) $\quad E_x^0 \left[e^{\lambda \mathcal{N}_1} \right] = 1 + (e^\lambda - 1) \sum_{n \geq 0} e^{\lambda n} P_x^0 [\mathcal{N}_1 > n].$

6. The large deviation estimate

Applying the strong Markov property and the invariance principle as in [**12**], (2.16) and below, we find that

(6.33) $$P_x^0[\mathcal{N}_1 > n] \leq (1-c)^n P_x^0[\mathcal{N}_1 > 0].$$

Choosing $\lambda > 0$ such that $e^\lambda(1-c) < 1$ with c as in (6.33), and coming back to (6.32), we see that for any $x \in S_{2N}$,

(6.34) $$E_x^0\left[e^{\lambda \mathcal{N}_1}\right] \leq 1 + c(\lambda) P_x^0[\mathcal{N}_1 > 0].$$

If we consider $|x_{d+1}| = 2N$, then we can apply the estimate (6.9) to

$$P_x^0[\mathcal{N}_1 > 0] = P_x^0\left[H_{C_{x_*}(L)}^X < \mathcal{D}_1\right]$$

with $A = C_{x_*}(L)$, $B = S_{4N}$, $a = 4N$ and then use the Green function estimates (6.10) for the numerator and (6.11) for the denominator of the right-hand side of (6.9), to obtain, for $N \geq c(\gamma')$,

$$P_x^0[\mathcal{N}_1 > 0] \leq cL^{d-1}N^{1-d}.$$

With (6.34), this yields, for any $x \in S_{2N}$ with $|x_{d+1}| = 2N$,

(6.35) $$E_x^0\left[e^{\lambda \mathcal{N}_1}\right] \leq 1 + c(\lambda)L^{d-1}N^{1-d}.$$

By (6.34), the first expectation on the right-hand side of (6.31) is bounded by a constant and with (6.35), the second expectation is bounded by $1 + c(\lambda)L^{d-1}N^{1-d}$. The estimate (6.30) follows and the proof of Lemma 6.6 is complete. □

We proceed with the proof of Theorem 6.1 when $\alpha \geq \frac{1}{d}$. For any $x \in S_{2N}$ and $m \geq 1$, we find

(6.36) $$P_x^0\left[U_{B_\infty(0,[N/4])} \leq \mathcal{D}_{[N^\beta]}\right] \leq$$
$$P_x^0\left[\text{for some } x_* \in B_\infty(0,[N/4]): \tilde{R}_m^{x_*} \leq \mathcal{D}_{[N^\beta]}\right] +$$
$$P_x^0\left[U_{B_\infty(0,[N/4])} \leq \mathcal{D}_{[N^\beta]}, \forall x_* \in B_\infty(0,[N/4]): \tilde{R}_m^{x_*} > \mathcal{D}_{[N^\beta]}\right]$$
$$\stackrel{\text{(def.)}}{=} P_1 + P_2.$$

Applying Lemma 6.6 to P_1, we obtain

(6.37) $$P_1 \leq cN^{d+1} \sup_{x_* \in B_\infty(0,[N/4])} P_x^0\left[\tilde{R}_m^{x_*} \leq \mathcal{D}_{[N^\beta]}\right]$$
$$\stackrel{(6.30)}{\leq} cN^{d+1} \exp\left\{cN^{1-d}L^{d-1}N^\beta - c'm\right\}.$$

In order order to bound P_2 in (6.36), we apply the geometric Lemmas 5.1 and 5.3. Together, they imply, for $N \geq c(\gamma,\gamma')$, the following

66   1. Disconnection of a discrete cylinder by a biased random walk

inclusions for the event $A_{\ldots,\ldots,\ldots}$ defined in (6.13):

(6.38) $\quad \left\{ U_{B_\infty(0,[N/4])} \leq \mathcal{D}_{[N^\beta]}, \text{ for all } x_* \in S_N : \tilde{R}_m^{x_*} > \mathcal{D}_{[N^\beta]} \right\} \subseteq$

$$\overset{(\text{Lemma 5.1})}{\subseteq} \bigcup_{\substack{x_* \in B_\infty(0,[N/4]), \\ C_{x_*}(L) \subseteq B_\infty(0,[N/4])}} \left\{ X([0, \tilde{D}_m^{x_*}]) \, \tfrac{1}{4}\text{-disc. } C_{x_*}(L) \right\}$$

$$\overset{(\text{Lemma 5.3})}{\subseteq} \bigcup_{\substack{x_* \in B_\infty(0,[N/4]), \\ C_{x_*}(L) \subseteq B_\infty(0,[N/4])}} A_{C_{x_*}(L), C_{x_*}(L)^{(L)}, l, c\left(\frac{L}{l}\right)^d, m}.$$

Since $1 \leq l \leq \frac{2L}{100}$ (cf. (5.1)) and $C_{x_*}(L)^{(\frac{2L}{10})} \subseteq C_{x_*}(L)^{(L)} \subseteq x_* + S_{2L}$ for $x_* \in B_\infty(0, [N/4])$ and $N \geq c(\gamma, \gamma')$, we can apply Lemma 6.5 with $a = 2L$ to obtain, for P_2 in (6.36),

$$P_2 \overset{(6.38)}{\leq} N^{d+1} \sup_{x_* \in B_\infty(0,[N/4])} P_x^0 \left[A_{C_{x_*}(L), C_{x_*}(L)^{(L)}, l, c\left(\frac{L}{l}\right)^d, m} \right]$$

$$\overset{(\text{Lemma 6.5, } a = 2L)}{\leq} N^{d+1} \exp\left\{ cm + cL^d l^{-d} \log N - c' L^{d-1} l^{-1} \right\}.$$

With (6.37) and (6.36), this estimate yields

(6.39) $\quad \sup_{x \in S_{2N}} P_x^0 \left[U_{B_\infty(0,[N/4])} \leq \mathcal{D}_{[N^\beta]} \right]$

$$\leq cN^{d+1} \exp\left\{ cN^{\beta - (d-1)(1-\gamma')} - c'm \right\}$$
$$+ N^{d+1} \exp\left\{ cm + cN^{d\gamma' - d\gamma} \log N - c'N^{(d-1)\gamma' - \gamma} \right\}.$$

In view of (5.1), we have $0 < \gamma < \gamma' < 1$, and provided $\beta - (d-1)(1-\gamma') < (d-1)\gamma' - \gamma$ and $d\gamma' - d\gamma < (d-1)\gamma' - \gamma$, i.e. if

(6.40) $\quad 0 < \gamma < \gamma' < 1, \quad \beta < d - 1 - \gamma, \quad \gamma' < (d-1)\gamma,$

then the right-hand side of (6.39) is bounded from above by

$$\exp\left\{ -cN^{(d-1)\gamma' - \gamma} \right\}$$

for $m_N = \left[c'' N^{\beta - (d-1)(1-\gamma')} \right]$ and $N \geq c(\gamma, \gamma')$, for a large enough constant $c'' > 0$. Hence, for γ, γ' satisfying (6.40), one has, for $N \geq c(\gamma, \gamma')$,

(6.41) $\quad \sup_{x \in S_{2N}} P_x^0 \left[U_{B_\infty(0,[N/4])} \leq \mathcal{D}_{[N^\beta]} \right] \leq \exp\left\{ -cN^{(d-1)\gamma' - \gamma} \right\}.$

For $0 < \beta < d - 1 - \frac{1}{d-1}$, $d \geq 3$, it is easy to check that the constraints (6.40) are satisfied by $\gamma_1 = \frac{1}{d-1} - \frac{\epsilon_1(d,\beta)}{2(d-1)}$ and $\gamma_1' = 1 - \epsilon_1(d, \beta)$ for $\epsilon_1(d, \beta) > 0$ small enough. Moreover, $(d-1)\gamma_1' - \gamma_1 = d - 1 - \frac{1}{d-1} - c\epsilon_1(d, \beta) = f(\alpha, \beta) - c\epsilon_1(d, \beta)$. By (6.41), this is enough to show (1.10), since we can make $\epsilon_1(d, \beta) > 0$ arbitrarily small.

6. The large deviation estimate

If, on the other hand, $d-1-\frac{1}{d-1} \leq \beta < d-1$, the constraints (6.40) are satisfied by $\gamma_2 = d-1-\beta-\frac{\epsilon_2(d,\beta)}{2(d-1)}$ and $\gamma_2' = (d-1)(d-1-\beta)-\epsilon_2(d,\beta)$ for $\epsilon_2(d,\beta) > 0$ sufficiently small and

$$(d-1)\gamma_2' - \gamma_2 = \left((d-1)^2 - 1\right)(d-1-\beta) - c\epsilon_2(d,\beta)$$
$$\stackrel{(6.1)}{=} f(\alpha,\beta) - c\epsilon_2(d,\beta),$$

which yields (1.10) for this range of β as well. This completes the proof of Theorem 6.1 for $\alpha \geq \frac{1}{d}$ and hence the proof of Theorem 6.1 altogether. □

REMARK 6.7. It is easy to see from Theorem 1.2 that the exponents in the upper and lower bounds on T_N for $\alpha < 1$ in (1.4) would match if one could show that the large deviation estimate (1.10) holds with the function f^* defined in (1.15). It may therefore be instructive to modify (1.10) by replacing the time $U_{B(\alpha)}$ by $\mathcal{U}$, defined as

$$\mathcal{U} = \inf\left\{n \geq 0 : X([0,n]) \supseteq \mathbb{T}_N^d \times \{0\}\right\}.$$

One can then show that f^* is indeed the correct exponent of the corresponding large deviation problem, in the following sense: For any α, $\beta > 0$, $0 < \xi_1 < f^*(\alpha,\beta) < \xi_2$, one has

$$(6.42) \quad \varlimsup_{N\to\infty} \frac{1}{N^{\xi_1}} \log \sup_{x \in S_{2[N^{d\alpha\wedge 1}]}} P_x^0\left[\mathcal{U} \leq \mathcal{D}_{[N^{\beta'}]}\right] < 0, \text{ for } 0 < \beta' < \beta,$$

as well as

$$(6.43) \quad \varlimsup_{N\to\infty} \frac{1}{N^{\xi_2}} \log \inf_{x \in S_{2[N^{d\alpha\wedge 1}]}} P_x^0\left[\mathcal{U} \leq \mathcal{D}_{[N^{\beta'}]}\right] = 0, \text{ for any } \beta' > \beta.$$

To show (6.42), one notes that standard estimates on one-dimensional random walk imply that the expected amount of time spent by the random walk X in $\mathbb{T}_N^d \times \{0\}$ during one excursion is of order $N^{d\alpha\wedge 1}$. With this information and the observation that $P^0[\mathcal{U} \leq \mathcal{D}_{[N^{\beta'}]}] \leq P^0[|X([0,\mathcal{D}_{[N^{\beta'}]}]) \cap \mathbb{T}_N^d \times \{0\}| \geq N^d]$, one can apply Khaśminskii's Lemma as in the proof of Lemma 6.5 to find the claimed upper bound. For (6.43), one can follow a similar route as in the derivation of the upper bounds on T_N. One can first establish Lemma 3.3 and hence the estimate (3.16) with ∞ replaced by $\mathcal{D}_1$, and then show that for S defined in (3.3) and a_N as in (3.22), $P^0[S_{[c_1 a_N]} \leq \mathcal{D}_1] \geq (1 - cN^{-d\alpha\wedge 1})^{c_1 a_N} \geq c\exp\left\{-c'N^{d-(d\alpha\wedge 1)}(\log N)^2\right\}$, where the first inequality follows essentially from standard estimates on one-dimensional random walk. This is enough for (6.43) with $\beta < d - (d\alpha \wedge 1)$. For $\beta \geq d - (d\alpha \wedge 1)$, one uses again that the expected number of visits to $\mathbb{T}_N^d \times \{0\}$ during one excursion is of order $N^{d\alpha\wedge 1}$, and finds that $P^0[S_{[c_1 a_N]} < \mathcal{D}_{[N^{\beta'}]}] \to 1$ as $N \to \infty$ for any $\beta' > \beta \geq d - (d\alpha \wedge 1)$. Using (3.16) for the second inequality, one deduces that for $N \geq c(\beta')$,

$$P_{\cdot}^0[\mathcal{C}_{\mathbb{T}_N^d}^V > [c_1 a_N] | S_{[c_1 a_N]} < \mathcal{D}_{[N^{\beta'}]}] \leq 2P_{\cdot}^0[\mathcal{C}_{\mathbb{T}_N^d}^V > [c_1 a_N]] \leq \frac{2}{N^{10}}, \text{ hence}$$

$$P_{\cdot}^0[\mathcal{U} \leq \mathcal{D}_{[N^{\beta'}]}] \geq P_{\cdot}^0[\mathcal{C}_{\mathbb{T}_N^d}^V \leq [c_1 a_N] | \bar{S}_{[c_1 a_N]} < \mathcal{D}_{[N^{\beta'}]}] P_{\cdot}^0[\bar{S}_{[c_1 a_N]} < \mathcal{D}_{[N^{\beta'}]}]$$
$$\xrightarrow{N \to \infty} 1,$$

thus (6.43) for $\beta \geq d - (d\alpha \wedge 1)$. Note that (6.43) and $\{\mathcal{U} \leq \mathcal{D}_{[N^{\beta'}]}\} \subseteq \{U_{B(\alpha)} \leq \mathcal{D}_{[N^{\beta'}]}\}$ together imply that, for any function f in the estimate (1.10), one has $f(\alpha, \beta') \leq f^*(\alpha, \beta)$ for any $\alpha, \beta > 0$, $\beta' > \beta$, so that $f(\alpha, \beta) \leq f^*(\alpha, \beta)$ whenever $f(\alpha, .)$ is right-continuous at β.

CHAPTER 2

Logarithmic components of the vacant set for random walk on a discrete torus

This work continues the investigation, initiated in a recent work by Benjamini and Sznitman, of percolative properties of the set of points not visited by a random walk on the discrete torus $(\mathbb{Z}/N\mathbb{Z})^d$ up to time uN^d in high dimension d. If $u > 0$ is chosen sufficiently small it has been shown that with overwhelming probability this vacant set contains a unique giant component containing segments of length $c_0 \log N$ for some constant $c_0 > 0$, and this component occupies a non-degenerate fraction of the total volume as N tends to infinity. Within the same setup, we investigate here the complement of the giant component in the vacant set and show that some components consist of segments of logarithmic size. In particular, this shows that the choice of a sufficiently large constant $c_0 > 0$ is crucial in the definition of the giant component.

1. Introduction

In a recent work by Benjamini and Sznitman [6], the authors consider a simple random walk on the d-dimensional integer torus $E = (\mathbb{Z}/N\mathbb{Z})^d$ for a sufficiently large dimension d and investigate properties of the set of points in the torus not visited by the walk after $[uN^d]$ steps for a sufficiently small parameter $u > 0$ and large N. Among other properties of this so-called vacant set, the authors of [6] find that for a suitably defined dimension-dependent constant $c_0 > 0$, there is a unique component of the vacant set containing segments of length at least $[c_0 \log N]$ with probability tending to 1 as N tends to infinity, provided $u > 0$ is chosen small enough. This component is referred to as the giant component. It is shown in [6] that with overwhelming probability, the giant component is at $|.|_\infty$-distance of at most N^β from any point and occupies at least a constant fraction γ of the total volume of the torus for arbitrary $\beta, \gamma \in (0,1)$, when $u > 0$ is chosen sufficiently small. One of the many natural questions that arise from the study of the giant component is whether there exist also other components in the vacant set containing segments of logarithmic size. In this work, we give an affirmative answer to this question. In particular, we show that for small $u > 0$, there exists some component consisting of a single segment of length $[c_1 \log N]$ for a dimension-dependent constant $c_1 > 0$ with probability tending to 1 as N tends to infinity.

In order to give a precise statement of this result, we introduce some notation and recall some results of [6]. Throughout this article, we denote the d-dimensional integer torus of side-length N by
$$E = (\mathbb{Z}/N\mathbb{Z})^d,$$
where the dimension $d \geq d_0$ is a sufficiently large integer (see (1.1)). E is equipped with the canonical graph structure, where any two vertices at Euclidean distance 1 are linked by an edge. We write P, or P_x, for $x \in E$, for the law on $E^\mathbb{N}$ endowed with the product σ-algebra $\mathcal{F}$, of the simple random walk on E started with the uniform distribution, or at x respectively. We let $(X_n)_{n \geq 0}$ stand for the canonical process on $E^\mathbb{N}$. By $X_{[s,t]}$, we denote the set of sites visited by the walk between times $[s]$ and $[t]$:
$$X_{[s,t]} = \left\{X_{[s]}, X_{[s]+1}, \ldots, X_{[t]}\right\}, \text{ for } s, t \geq 0.$$
We use the notation $e_1, \ldots, e_d$ for the canonical basis of $\mathbb{R}^d$, and denote the segment of length $l \geq 0$ in the e_i-direction at $x \in E$ by
$$[x, x + le_i] = E \cap \{x + \lambda l e_i : \lambda \in [0,1]\},$$
where the addition is naturally understood as addition modulo N. The authors of [6] introduce a dimension-dependent constant $c_0 > 0$ (cf. [6], (2.47)) and for any $\beta \in (0,1)$ define an event $\mathcal{G}_{\beta,t}$ for $t \geq 0$ (cf. [6], (2.52) and Corollary 2.6 in [6]), on which there exists a unique component O of $E \setminus X_{[0,t]}$ containing any segment in $E \setminus X_{[0,t]}$ of the form $[x, x + [c_0 \log N]e_i]$, $i = 1, \ldots, d$, and such that O is at an $|.|_\infty$-distance of at most N^β from any point in E. This unique component is referred to as the *giant component*. As in [6], we consider dimensions $d \geq d_0$, with d_0 defined as the smallest integer $d_0 \geq 5$ such that

$$(1.1) \qquad 49\left(\frac{2}{d} + \left(1 - \frac{2}{d}\right)q(d-2)\right) < 1 \quad \text{for any } d \geq d_0,$$

where $q(d)$ denotes the probability that the simple random walk on $\mathbb{Z}^d$ returns to its starting point. Note that d_0 is well-defined, since $q(d) \downarrow 0$ as $d \to \infty$ (see [22], (5.4), for precise asymptotics of $q(d)$). Among other properties of the vacant set, it is shown in [6], Corollary 4.6, that for any dimension $d \geq d_0$ and any $\beta, \gamma \in (0,1)$,

$$(1.2) \qquad \lim_N P\left[\mathcal{G}_{\beta,uN^d} \cap \left\{\frac{|O|}{N^d} \geq \gamma\right\}\right] = 1, \quad \text{for small } u > 0.$$

Our main result is:

THEOREM 1.1. *($d \geq d_0$) For any sufficiently small $u > 0$, the vacant set left by the random walk on $(\mathbb{Z}/N\mathbb{Z})^d$ up to time uN^d contains some segment of length*

$$(1.3) \qquad l = [c_1 \log N] \stackrel{(\text{def.})}{=} \left[(300d \log(2d))^{-1} \log N\right],$$

1. Introduction

which does not belong to the giant component with probability tending to 1 as $N \to \infty$. That is, for any $\beta \in (0,1)$,

(1.4) $$\lim_N P\left[\mathcal{G}_{\beta, uN^d} \cap \left(\bigcup_{x \in E} \{[x, x+le_1] \subseteq E \setminus (X_{[0, uN^d]} \cup O)\}\right)\right] = 1,$$

for small $u > 0$.

We now comment on the strategy of the proof of Theorem 1.1. We show that for l as in (1.3), for some $\nu > 0$ and $u > 0$ chosen sufficiently small,

(1.5)   the vacant set at time $\left[N^{2-\frac{1}{10}}\right]$ contains at least $[N^\nu]$ components consisting of a single segment of length l (cf. Section 3),

(1.6)   with high probability some of these segments remain unvisited until time $[uN^d]$ (cf. Section 5).

Note that these logarithmic components are distinct from the giant component with overwhelming probability in view of (1.2).

Let us explain the main ideas in the proofs of the claims (1.5) and (1.6). The argument showing (1.5) consists of two steps. The first step is Lemma 3.2, which proves that with high probability, at any two times until $\left[N^{2-\frac{1}{10}}\right]$ separated by at least $\left[N^{\frac{4}{3}}\right]$, the random walk is at distinct locations. Here, the fact that $d \geq 5$ plays an important role.

In the second step, we partition the time interval $\left[0, \left[N^{2-\frac{1}{10}}\right]\right]$ into subintervals of length $\left[N^{\frac{4}{3}+\frac{1}{100}}\right] > \left[N^{\frac{4}{3}}\right]$. We show in Lemma 3.3 that with high probability, there are at least $[N^\nu]$ such subintervals during which the following phenomenon occurs: the random walk visits every point on the boundary of an unvisited segment of length l without hitting the segment itself, and thereafter also does not visit the segment for a time longer than $\left[N^{\frac{4}{3}}\right]$. It then follows with the help of the previous Lemma 3.2 that the random walk does not visit the surrounded segments at all. Similarly, the segments surrounded in the $[N^\nu]$ different subintervals are seen to be distinct, and claim (1.5) is shown (cf. Lemma 3.4). The proof of Lemma 3.3 uses a result on the ubiquity of segments of logarithmic size in the vacant set from [6]. From this ubiquity result, we know that for any $\beta > 0$, with overwhelming probability, there is a segment of length l in the vacant set left until the beginning of every considered subinterval (in fact even until $[uN^d]$ for small $u > 0$) in the N^β-neighborhood of any point. Hence, to show Lemma 3.3, it essentially suffices to find a lower bound on the probability that for some $\beta > 0$, the random walk surrounds, but does not visit, a fixed segment in the N^β-neighborhood of its starting point until time $\left[N^{\frac{4}{3}+\frac{1}{100}}/2\right]$ and does not visit the same segment until time $\left[N^{\frac{4}{3}+\frac{1}{100}}\right] > \left[N^{\frac{4}{3}+\frac{1}{100}}/2\right] + \left[N^{\frac{4}{3}}\right]$.

The rough idea behind the proof of claim (1.6) is to use a lower bound on the probability that one fixed segment of length l survives (i.e. remains unvisited) for a time of at least $[uN^d]$. With estimates on hitting probabilities mentioned in Section 2, it can be shown that this probability is at least $e^{-\text{const } ul}$. Since this is much larger than $\frac{1}{[N^\nu]}$ for $u > 0$ sufficiently small, cf. (1.3), it should be expected that with high probability, at least one of the $[N^\nu]$ unvisited segments survives until time $[uN^d]$. This conclusion does not follow immediately, because of the dependence between the events that different segments survive. However, the desired conclusion does follow by an application of a technique, developed in [6], for bounding the variance of the total number of segments which survive.

The article is organized as follows:

Section 2 contains some estimates on hitting probabilities and exit times recurrently used throughout this work. In Section 3, we prove claim (1.5). In Section 4, we prove a crucial ingredient for the derivation of claim (1.6). In Section 5, we prove (1.6) and conclude that these two ingredients do yield Theorem 1.1.

Finally, we use the following convention concerning constants: Throughout the text, c or c' denote positive constants which only depend on the dimension d, with values changing from place to place. The numbered constants c_0, c_1, c_2, c_3, c_4 are fixed and refer to their first place of appearance in the text.

Acknowledgments. The author is grateful to Alain-Sol Sznitman for proposing the problem and for helpful advice.

2. Some definitions and useful results

In this section, we introduce some more standard notation and some preliminary estimates on hitting probabilities and exit and return times to be frequently used later on. By $(\mathcal{F}_n)_{n\geq 0}$ and $(\theta_n)_{n\geq 0}$ we denote the canonical filtration and shift operators on $E^{\mathbb{N}}$. For any set $A \subseteq E$, we often consider the entrance time H_A and the exit time T_A, defined as

$$H_A = \inf\{n \geq 0 : X_n \in A\}, \text{ and}$$
$$T_A = \inf\{n \geq 0 : X_n \notin A\}.$$

For any set $B \subsetneq E$, we denote the Green function of the random walk killed when exiting B as

$$(2.1) \qquad g^B(x,y) = E_x\left[\sum_{n=0}^\infty \mathbf{1}\{X_n = y, n < T_B\}\right].$$

2. Some definitions and useful results

We write $|.|_\infty$ for the l_∞-distance on E, $B(x,r)$ for the $|.|_\infty$-closed ball of radius $r > 0$ centered at $x \in E$, and denote the induced mutual distance of subsets A, B of E with

$$d(A,B) = \inf\{|x - y|_\infty : x \in A, y \in B\}.$$

For any set $A \subseteq E$, the boundary ∂A of A is defined as the set of points in $E \setminus A$ having neighbors in A and the number of points in A is denoted by $|A|$. For sequences a_N and b_N, we write $a_N \ll b_N$ to mean that a_N/b_N tends to 0 as N tends to infinity.

Throughout the proof, we often use the following estimate on hitting probabilities:

LEMMA 2.1. $(d \geq 1, A \subseteq B \subsetneq E, x \in B)$

$$(2.2) \qquad \frac{\sum_{y \in A} g^B(x,y)}{\sup_{y \in A} \sum_{y' \in A} g^B(y,y')} \leq P_x[H_A \leq T_B] \leq \frac{\sum_{y \in A} g^B(x,y)}{\inf_{y \in A} \sum_{y' \in A} g^B(y,y')}.$$

PROOF. Apply the strong Markov property at H_A to

$$\sum_{y \in A} g^B(x,y) = E_x\left[\{H_A \leq T_B\}, \left(\sum_{y \in A} g^B(X_0, y)\right) \circ \theta_{H_A}\right].$$

$\square$

Moreover, we use the following exit-time estimates:

LEMMA 2.2. $(1 \leq a, b < \frac{N}{2}, x \in E)$

$$(2.3) \qquad P_x\left[T_{B(0,a)} \geq b^2\right] \leq c e^{-c'\left(\frac{b}{a}\right)^2},$$

$$(2.4) \qquad P_0\left[T_{B(0,b)} \leq a^2\right] \leq c e^{-c'\frac{b}{a}}.$$

PROOF. We may assume that $2a \leq b$, for otherwise there is nothing to prove. To show (2.3), one uses the Chebychev inequality with $\lambda > 0$ and obtains

$$P_x\left[T_{B(0,a)} \geq b^2\right] \leq E_x\left[\exp\left\{\frac{\lambda}{a^2} T_{B(0,a)}\right\}\right] e^{-\lambda\left(\frac{b}{a}\right)^2}.$$

By Khaśminskii's Lemma (see [**33**], Lemma 1.1, p. 292, and also [**18**]), this last expectation is bounded from above by 2 for a certain constant $\lambda > 0$, and (2.3) follows. As for (2.4), we define the stopping times $(U_n)_{n \geq 1}$ as the times of successive displacements of the walk at distance a, i.e.

$$U_1 = \inf\{n \geq 0 : |X_n - X_0|_\infty \geq a\}, \text{ and for } n \geq 2,$$
$$U_n = U_1 \circ \theta_{U_{n-1}} + U_{n-1}.$$

Since $b \geq \left[\frac{b}{a}\right]a$, one has $T_{B(0,b)} \geq U_{\left[\frac{b}{a}\right]}$ P_0-a.s., hence by the Chebychev inequality and the strong Markov property applied inductively at the times $U_{[\frac{b}{a}]-1}, \ldots, U_1$,

$$P_0\left[T_{B(0,b)} \leq a^2\right] \leq eE_0\left[\exp\left\{-\frac{1}{a^2}U_{[\frac{b}{a}]}\right\}\right]$$

$$\stackrel{\text{(Markov)}}{\leq} e\left(E_0\left[\exp\left\{-\frac{1}{a^2}U_1\right\}\right]\right)^{[\frac{b}{a}]}.$$

By the invariance principle, the last expectation is bounded from above by $1 - c$ for some constant $c > 0$, from which (2.4) follows. □

The following positive constants remain fixed throughout the article,

(2.5) $\quad \beta_0 = \dfrac{1}{3(d-2)} < \alpha_0 = \dfrac{4}{3} < \beta_1 = \dfrac{4}{3} + \dfrac{1}{100} < \alpha_1 = 2 - \dfrac{1}{10}$,

as do the quantities

(2.6) $\quad b_0 = [N^{\beta_0}] \ll a_0 = [N^{\alpha_0}] \ll b_1 = [N^{\beta_1}] \ll a_1 = [N^{\alpha_1}]$.

We are now ready to begin the proof of the two crucial claims (1.5) and (1.6), starting with (1.5).

3. Profusion of logarithmic components until time a_1

In this section, we show the claim (1.5). To this end, we define the $\mathcal{F}_{[t]}$-measurable random subset $\mathcal{J}_t$ of E for $t \geq 0$, as the set of all $x \in E$ such that the segment $[x, x + le_1]$ forms a component of the vacant set left until time $[t]$, where l was defined in (1.3):

(3.1) $\quad \mathcal{J}_t = \{x \in E : X_{[0,t]} \supseteq \partial[x, x + le_1]$

$\text{and } X_{[0,t]} \cap [x, x + le_1] = \emptyset\}$.

We then show that for small $\nu > 0$, at least $[N^\nu]$ segments of length l occur as components in the vacant set until time a_1 with overwhelming probability:

PROPOSITION 3.1. $(d \geq 5, a_1$ as in (2.6), l as in (1.3)) For small $\nu > 0$,

(3.2) $\quad \lim_N P\left[|\mathcal{J}_{a_1}| \geq [N^\nu]\right] = 1$.

PROOF. The proof of Proposition 3.1 will be split into Lemmas 3.2, 3.3 and 3.4, which we now state. Lemma 3.2 asserts that when $d \geq 5$, on an event of probability tending to 1 as N tends to infinity, $X_I \cap X_J = \emptyset$, for all subintervals I, J of $[0, a_1]$ with mutual distance at least a_0.

LEMMA 3.2. $(d \geq 5)$

(3.3) $\quad \lim_N P\left[\bigcap_{n=0}^{a_1-a_0}\{X_{[0,n]} \cap X_{[n+a_0, a_1]} = \emptyset\}\right] = 1$.

3. Profusion of logarithmic components until time a_1

We then consider the $[a_1/b_1]$ subintervals
$$[(i-1)b_1, ib_1], i = 1, \ldots [a_1/b_1],$$
of the interval $[0, a_1]$, each of length b_1, larger than a_0, cf. (2.6). By $\mathcal{A}_{i,S}$, $S \subseteq E$, we denote the event that, during the first half of the i-th time interval, the random walk produces a component consisting of a segment of length l (cf. (1.3)) at some point $x \in S$, and does not visit the same component until the end of the i-th time interval:

(3.4) $\quad \mathcal{A}_{i,S} = \bigcup_{x \in S} \left(\left\{ X_{[(i-1)b_1, (i-1)b_1+b_1/2]} \supseteq \partial[x, x+le_1] \right\} \cap \right.$
$$\left. \left\{ X_{[0, ib_1]} \cap [x, x+le_1] = \emptyset \right\} \right) \in \mathcal{F}_{ib_1},$$

for $i = 1, \ldots, [a_1/b_1]$. For $S \subseteq E$, the random subset $\mathcal{I}_S$ of
$$\{1, \ldots, [a_1/b_1]\}$$
is then defined as the set of indices i for which $\mathcal{A}_{i,S}$ occurs, i.e.

(3.5) $\quad \mathcal{I}_S = \{i \in \{1, \ldots, [a_1/b_1]\} : \mathcal{A}_{i,S} \text{ occurs}\}.$

The next lemma then asserts that at least $[N^\nu]$ of the events $\mathcal{A}_{i,E}$, $i = 1, \ldots, [a_1/b_1]$, occur.

LEMMA 3.3. $(d \geq 4)$ For small $\nu > 0$,

(3.6) $\quad \lim_N P\left[|\mathcal{I}_E| \geq [N^\nu]\right] = 1.$

Finally, Lemma 3.4 shows that Lemmas 3.2 and 3.3 together do yield Proposition 3.1.

LEMMA 3.4. $(d \geq 2, \nu > 0, N \geq c)$

(3.7) $\quad \{|\mathcal{I}_E| \geq [N^\nu]\} \cap \bigcap_{n=0}^{a_1 - a_0} \{X_{[0,n]} \cap X_{[n+a_0, a_1]} = \emptyset\}$
$$\subseteq \{|\mathcal{J}_{a_1}| \geq [N^\nu]\}.$$

We now prove these three Lemmas.

PROOF OF LEMMA 3.2. We start by observing that by the simple Markov property and translation invariance, the probability of the complement of the event in (3.3) is bounded by

(3.8) $\quad P\left[\bigcup_{\substack{n,m \in [0, a_1] \\ m \geq n + a_0}} \{X_n = X_m\}\right] \leq \sum_{n=0}^{a_1} P\left[\bigcup_{m \in [n+a_0, n+a_1]} \{X_n = X_m\}\right]$
$$= (a_1 + 1) P_0 \left[H_{\{0\}} \circ \theta_{a_0} + a_0 \leq a_1 \right].$$

The remaining task is to find an upper bound on this last probability via the exit-time estimates (2.3) and (2.4). We put $a_* = \left[N^{\frac{a_0}{2} - \frac{1}{100}}\right] = \left[N^{\frac{2}{3} - \frac{1}{100}}\right]$. Note that then $a_*^2 \ll a_0$ and $a_1 \ll N^2$. By the exit-time estimates (2.3) and (2.4), we can therefore assume that the random walk exits the ball $B(0, a_*)$ before time a_0, but remains in $B(0, \frac{N}{4})$ until time a_1. More precisely, one has

(3.9)
$$P_0\left[H_{\{0\}} \circ \theta_{a_0} + a_0 \leq a_1\right]$$
$$\leq P_0\left[H_{\{0\}} \circ \theta_{a_0} + a_0 \leq a_1, T_{B(0,a_*)} \leq a_0, T_{B(0,\frac{N}{4})} > a_1\right]$$
$$+ P_0\left[T_{B(0,a_*)} > a_0\right] + P_0\left[T_{B(0,\frac{N}{4})} \leq a_1\right]$$
$$= P_1 + P_2 + P_3,$$

where P_1, P_2 and P_3 is abbreviated notation for three terms in the previous line. By the exit-time estimate (2.3) applied with $a = a_*$ and $b = \sqrt{a_0}$, one has

(3.10) $\quad P_2 = P_0\left[T_{B(0,a_*)} > a_0\right] \leq ce^{-c'\frac{a_0}{a_*^2}} \leq ce^{-c'N^{\frac{1}{50}}}.$

Moreover, the estimate (2.4) with $a = \sqrt{a_1}$ and $b = \frac{N}{4}$ implies that

(3.11) $\quad P_3 = P_0\left[T_{B(0,\frac{N}{4})} \leq a_1\right] \leq ce^{-c'\frac{N}{\sqrt{a_1}}} \leq ce^{-c'N^{\frac{1}{20}}}.$

It thus remains to bound P_1. We obtain by the strong Markov property applied at time $T_{B(0,a_*)}$, that

(3.12) $\quad P_1 \leq P_0\left[H_{\{0\}} \circ \theta_{T_{B(0,a_*)}} + T_{B(0,a_*)} < T_{B(0,\frac{N}{4})}\right]$
$$\stackrel{\text{(Markov)}}{\leq} \sup_{x \in E: |x|_\infty = a_* + 1} P_x\left[H_{\{0\}} \leq T_{B(0,\frac{N}{4})}\right]$$

The standard Green function estimate from [19], Theorem 1.5.4. implies that for any $x \in E$ with $|x|_\infty = a_* + 1$,

$$P_x\left[H_{\{0\}} \leq T_{B(0,\frac{N}{4})}\right] \stackrel{(2.1)}{\leq} g^{B(0,\frac{N}{4})}(x,0) \leq ca_*^{-(d-2)} \leq cN^{-(d-2)\left(\frac{\alpha_0}{2} - \frac{1}{100}\right)}.$$

Inserted into (3.12), this yields

(3.13) $\quad P_1 \leq cN^{-(d-2)\left(\frac{\alpha_0}{2} - \frac{1}{100}\right)}.$

Substituting the bounds (3.10), (3.11) and (3.13) into (3.9), one then finds that

$$P_0\left[H_{\{0\}} \circ \theta_{a_0} + a_0 \leq a_1\right] \leq cN^{-(d-2)\left(\frac{\alpha_0}{2} - \frac{1}{100}\right)}.$$

Inserting this estimate into (3.8), one finally obtains

(3.14) $\quad P\left[\bigcup_{\substack{n,m \in [0,a_1] \\ m \geq n + a_0}} \{X_n = X_m\}\right] \leq ca_1 N^{-(d-2)\left(\frac{\alpha_0}{2} - \frac{1}{100}\right)}$
$$\leq cN^{\alpha_1 - (d-2)\left(\frac{\alpha_0}{2} - \frac{1}{100}\right)}.$$

3. Profusion of logarithmic components until time a_1

Since $d - 2 \geq 3$, we have

$$\alpha_1 - (d-2)\left(\frac{\alpha_0}{2} - \frac{1}{100}\right) \leq 2 - \frac{1}{10} - 3\left(\frac{2}{3} - \frac{1}{100}\right) = -\frac{7}{100} < 0,$$

and the proof of Lemma 3.2 is complete with (3.14). □

PROOF OF LEMMA 3.3. The following result on the ubiquity of segments of logarithmic size from [6] will be used: Define for any constants $K > 0$, $0 < \beta < 1$ and time $t \geq 0$, the event

$$(3.15) \quad \mathcal{V}_{K,\beta,t} = \Big\{ \text{for all } x \in E,\ 1 \leq j \leq d,\ \text{for some } 0 \leq m < N^\beta,$$
$$X_{[0,t]} \cap \{x + (m + [0, [K \log N]])\, e_j\} = \emptyset \Big\}.$$

Then for dimension $d \geq 4$ and some constant $c > 0$, one has

$$(3.16) \quad \limsup_N \frac{1}{N^c} \log P\left[\mathcal{V}^c_{c_1,\beta_0,uN^d}\right] < 0, \quad \text{for small } u > 0,$$

see the end of the proof of Theorem 1.2 in [6] and note the bounds (1.11), (1.49), (1.56) in [6]. With this last estimate we will be able to assume that at the beginning of every time interval $[(i-1)b_1, ib_1]$, $i = 1, \ldots, [a_1/b_1]$, there is an unvisited segment of length l in the b_0-neighborhood of the current position of the random walk. This will reduce the proof of Lemma 3.3 to the derivation of a lower bound on $P_0\left[\mathcal{A}_{1,\{x\}}\right]$ for an x in the b_0-neighborhood of 0.

We denote with $\mathbb{I}$ the set of indices, i.e. $\mathbb{I} = \{1, \ldots, [a_1/b_1]\}$. A rough counting argument yields the following bound on the probability of the complement of the event in (3.6):

$$(3.17) \quad P\left[|\mathcal{I}_E| < [N^\nu]\right] \leq \sum_{\substack{I \subseteq \mathbb{I} \\ |I| \geq |\mathbb{I}| - [N^\nu]}} P\left[\mathcal{I}^c_E \supseteq I\right]$$
$$\leq e^{cN^\nu \log N} \sup_{\substack{I \subseteq \mathbb{I} \\ |I| \geq |\mathbb{I}| - N^\nu}} P\left[\mathcal{I}^c_E \supseteq I\right].$$

For any set I considered in the last supremum, we label its elements in increasing order as $1 \leq i_1 < \ldots < i_{|I|}$. Note that the events $\mathcal{V}_{c_1,\beta_0,t}$ defined in (3.15) decrease with t. Applying (3.16), one obtains that

$$(3.18) \quad P\left[\mathcal{I}^c_E \supseteq I\right] \leq P\left[\{\mathcal{I}^c_E \supseteq I\} \cap \mathcal{V}_{c_1,\beta_0,a_1}\right] + ce^{-N^{c'}}.$$

Again with monotonicity of $\mathcal{V}_{c_1,\beta_0,t}$ in t, one finds

$$(3.19) \quad P\left[\{\mathcal{I}^c_E \supseteq I\} \cap \mathcal{V}_{c_1,\beta_0,a_1}\right]$$
$$\leq P\left[\bigcap_{i \in I \setminus \{i_{|I|}\}} \mathcal{A}^c_{i,E} \cap \mathcal{V}_{c_1,\beta_0,(i_{|I|}-1)b_1} \cap \mathcal{A}^c_{i_{|I|},E}\right].$$

We now claim that for any event $\mathcal{B} \in \mathcal{F}_{(i-1)b_1}$, $i \in \mathbb{I}$, such that $\mathcal{B} \subseteq \mathcal{V}_{c_1,\beta_0,(i-1)b_1}$, we have

(3.20) $\qquad P[\mathcal{A}_{i,E} \cap \mathcal{B}] \geq c b_0^{-(d-2)} N^{-\frac{1}{100}} P[\mathcal{B}]$, for $N \geq c'$.

Before proving (3.20), we note that if one uses (3.20) in (3.19) with $i = i_{|I|}$ and $\mathcal{B} = \bigcap_{i \in I \setminus \{i_{|I|}\}} \mathcal{A}_{i,E}^c \cap \mathcal{V}_{c_1,\beta_0,(i_{|I|}-1)b_1} \in \mathcal{F}_{(i_{|I|}-1)b_1}$, one obtains for $N \geq c$,

$$P\left[\bigcap_{i \in I} \mathcal{A}_{i,E}^c \cap \mathcal{V}_{c_1,\beta_0,a_1}\right] \leq P\left[\bigcap_{i \in I \setminus \{i_{|I|}\}} \mathcal{A}_{i,E}^c \cap \mathcal{V}_{c_1,\beta_0,(i_{|I|}-1)b_1}\right]$$
$$\times \left(1 - c' b_0^{-(d-2)} N^{-\frac{1}{100}}\right),$$

and proceeding inductively, one has for $0 < \nu < (\alpha_1 - \beta_1)/2$ (cf. (2.5)) and $N \geq c$,

(3.21) $\qquad P[\{\mathcal{I}_E^c \supseteq I\} \cap \mathcal{V}_{c_1,\beta_0,a_1}] \leq \left(1 - c' b_0^{-(d-2)} N^{-\frac{1}{100}}\right)^{|I|}$
$$\leq \exp\left\{-c' N^{-(d-2)\beta_0 - \frac{1}{100} + \alpha_1 - \beta_1}\right\}$$
$$\stackrel{(2.5)}{\leq} \exp\left\{-c' N^{\frac{1}{6}}\right\}.$$

As a result, (3.17), (3.18) and (3.21) together yield for $0 < \nu < (\alpha_1 - \beta_1)/2$ and $N \geq c$,

$$P[|\mathcal{I}_E| < [N^\nu]] \leq \exp\left\{N^\nu \log N - c' N^{\frac{1}{6}}\right\} + c'' \exp\left\{N^\nu \log N - N^{c'}\right\},$$

hence (3.6). It therefore only remains to show (3.20). To this end, we first find a suitable unvisited segment of length l to be surrounded during the i-th time interval. We thus define the $\mathcal{F}_{(i-1)b_1}$-measurable random subsets $(\mathcal{K}_S)_{S \subseteq E}$ of E of points $x \in S \subseteq E$ such that the segment of length l at site $X_{(i-1)b_1} + x$ is vacant at time $(i-1)b_1$:

$$\mathcal{K}_S = \left\{x \in S : X_{[0,(i-1)b_1]} \cap \left(X_{(i-1)b_1} + x + [0, le_1]\right) = \emptyset\right\}.$$

For $N \geq c$, on the event $\mathcal{V}_{c_1,\beta_0,(i-1)b_1}$, for any $y \in E$ there is an integer $0 \leq m \leq b_0$ such that the segment $y + me_1 + [0, le_1]$ is contained in the vacant set left until time $(i-1)b_1$. This implies in particular that with $y = X_{(i-1)b_1}$ (and necessarily $m > 0$):

$$\mathcal{V}_{c_1,\beta_0,(i-1)b_1} \subseteq \{\mathcal{K}_{[e_1,b_0e_1]} \neq \emptyset\}.$$

Since the event $\mathcal{B}$ in (3.20) is a subset of $\mathcal{V}_{c_1,\beta_0,(i-1)b_1}$, it follows that

(3.22) $\qquad P[\mathcal{A}_{i,E} \cap \mathcal{B}] = P\left[\mathcal{B} \cap \{\mathcal{K}_{[e_1,b_0e_1]} \neq \emptyset\} \cap \mathcal{A}_{i,E}\right]$
$$= \sum_{\substack{S \subseteq [e_1,b_0e_1], \\ S \neq \emptyset}} P\left[\mathcal{B} \cap \{\mathcal{K}_{[e_1,b_0e_1]} = S\} \cap \mathcal{A}_{i,E}\right].$$

3. Profusion of logarithmic components until time a_1

Observe that for any $S \subseteq [e_1, b_0 e_1]$, $\{\mathcal{K}_{[e_1,b_0e_1]} = S\} \cap \theta^{-1}_{(i-1)b_1}\mathcal{A}_{1,S} \subseteq \mathcal{A}_{i,S} \subseteq \mathcal{A}_{i,E}$, so it follows from (3.22) that

$$P[\mathcal{A}_{i,E} \cap \mathcal{B}] \geq \sum_{\substack{S \subseteq [e_1,b_0e_1],\\ S \neq \emptyset}} P\left[\mathcal{B} \cap \{\mathcal{K}_{[e_1,b_0e_1]} = S\} \cap \theta^{-1}_{(i-1)b_1}\mathcal{A}_{1,S}\right].$$

Note that $\mathcal{K}_{[e_1,b_0e_1]}$ and $\mathcal{B}$ are both $\mathcal{F}_{(i-1)b_1}$-measurable. Applying the simple Markov property at time $(i-1)b_1$ to the probability in this last expression and using translation invariance, it follows that

$$(3.23) \qquad P[\mathcal{A}_{i,E} \cap \mathcal{B}] \geq \inf_{\substack{S \subseteq [e_1,b_0e_1]\\ S \neq \emptyset}} P_0[\mathcal{A}_{1,S}] P[\mathcal{B}]$$

$$\geq \inf_{x \in [e_1,b_0e_1]} P_0\left[\mathcal{A}_{1,\{x\}}\right] P[\mathcal{B}].$$

In the remainder of this proof, we find a lower bound on

$$\inf_{x \in [e_1,b_0e_1]} P_0\left[\mathcal{A}_{1,\{x\}}\right]$$

in three steps. First, for arbitrary $x \in [e_1, b_0 e_1]$, we bound from below the probability that the random walk reaches the boundary $\partial[x, x + l e_1]$ within time at most $b_1/4$. Next, we estimate the probability that the random walk, once it has reached $\partial[x, x + l e_1]$, covers $\partial[x, x + l e_1]$ in $[3dl] \ll b_1/4$ steps. And finally, we find a lower bound on the probability that the random walk starting from $\partial[x, x + l e_1]$ does not visit the segment $[x, x + l e_1]$ during a time interval of length b_1. With this program in mind, note that for $x \in [e_1, b_0 e_1]$ and $N \geq c'$, one has

$$\mathcal{A}_{1,\{x\}} \supseteq \left\{H_{\partial[x,x+le_1]} \leq \frac{1}{4}b_1\right\} \cap \left\{(X \circ \theta_{H_{\partial[x,x+le_1]}})_{[0,[3dl]]} = \partial[x, x+le_1]\right\}$$

$$\cap \left\{(X \circ \theta_{H_{\partial[x,x+le_1]}+[3dl]})_{[0,b_1]} \cap [x, x+le_1] = \emptyset\right\}, \quad P_0\text{-a.s.}$$

By the strong Markov property, applied at time $H_{\partial[x,x+le_1]} + [3dl]$, then at time $H_{\partial[x,x+le_1]}$, and translation invariance, one can thus infer that

$$(3.24) \qquad \inf_{x \in [e_1,b_0e_1]} P_0\left[\mathcal{A}_{1,\{x\}}\right] \geq \inf_{x \in E: |x|_\infty \leq b_0} P_0\left[H_{\partial[x,x+le_1]} \leq \frac{1}{4}b_1\right]$$

$$\times \inf_{y \in \partial[0,le_1]} P_y\left[X_{[0,[3dl]]} = \partial[0, le_1]\right]$$

$$\times \inf_{y \in \partial[0,le_1]} P_y\left[X_{[0,b_1]} \cap [0, le_1] = \emptyset\right]$$

$$\stackrel{(\text{def.})}{=} L_1 L_2 L_3.$$

We now bound each of the above factors from below. Beginning with L_1, we fix $x \in E$ such that $|x|_\infty \leq b_0$ and define $b_* = \left[N^{\frac{1}{2}(\beta_1 - \frac{1}{100})}\right] = \left[N^{\frac{2}{3}}\right]$ (so that $b_0 \ll b_*$ and $b_*^2 \ll b_1$). We then observe that

$$P_0\left[H_{\partial[x,x+le_1]} \leq \frac{1}{4}b_1\right] \geq P_0\left[H_{\partial[x,x+le_1]} \leq T_{B(0,b_*)}\right] - P_0\left[T_{B(0,b_*)} \geq \frac{1}{4}b_1\right].$$

80   2. Vacant set for random walk on a discrete torus

With (2.3), where $a = b_*$ and $b = \sqrt{\frac{b_1}{4}}$, we infer with (2.5) that

(3.25) $\quad P_0\left[H_{\partial[x,x+le_1]} \leq \frac{1}{4}b_1\right] \geq P_0\left[H_{\partial[x,x+le_1]} \leq T_{B(0,b_*)}\right]$

$$- c\exp\left\{-c'N^{\frac{1}{100}}\right\}.$$

We then use the left-hand estimate of (2.2) to find that

$$P_0\left[H_{\partial[x,x+le_1]} \leq T_{B(0,b_*)}\right] \geq \frac{\sum_{y \in \partial[x,x+le_1]} g^{B(0,b_*)}(0,y)}{\sup_{y \in \partial[x,x+le_1]} \sum_{y' \in \partial[x,x+le_1]} g^{B(0,b_*)}(y,y')}.$$

With the Green function estimate of [**19**], Proposition 1.5.9 (for the numerator) and transience of the simple random walk in dimension $d-1$ (for the denominator), the right-hand side is bounded from below by $clb_0^{-(d-2)}$. With (3.25), this implies that for $N \geq c$,

(3.26) $\quad L_1 \geq c' l b_0^{-(d-2)}.$

The lower bound we need on L_2 in (3.24) is straightforward: We simply calculate the probability that the random walk follows a suitable fixed path in $\partial[0, le_1]$, starting at $y \in \partial[0, le_1]$ and covering $\partial[0, le_1]$ in at most $d(2l+8) \leq 3dl$ steps (for $N \geq c'$). Such a path can for instance be found by considering the paths $\mathcal{P}_i$, $i = 2, \ldots, d$, surrounding the segment $[0, le_1]$ in the (e_1, e_i)-hyperplane, i.e.

$$\mathcal{P}_i = (-1e_1 + 0e_i, -1e_1 + 1e_i, 0e_1 + 1e_i, 1e_1 + 1e_i, \ldots, (l+1)e_1 + 1e_i,$$
$$(l+1)e_1 + 0e_i, (l+1)e_1 - 1e_i, le_1 - 1e_i, \ldots, -1e_1 - 1e_i, -1e_1 + 0e_i),$$

$i = 2, \ldots, d$. The paths $\mathcal{P}_i$ visit only points in $\partial[0, le_1]$ and their concatenation forms a path starting at $-e_1$ and covering $\partial[0, le_1]$ in $(d-1)(2l+8)$ steps. Finally, any starting point $y \in \partial[0, le_1]$ is linked to $-e_1$ in $\leq 2l + 8$ steps via one of the paths $\mathcal{P}_i$. Therefore, we have

(3.27) $\quad L_2 \geq \left(\frac{1}{2d}\right)^{3dl} = e^{-(3d\log 2d)l} \overset{(1.3)}{\geq} N^{-\frac{1}{100}}.$

For L_3 in (3.24), we note that for any $y \in \partial[0, le_1]$,

(3.28) $\quad P_y\left[X_{[0,b_1]} \cap [0, le_1] = \emptyset\right] \geq P_y\left[T_{B(0,\frac{N}{4})} < H_{[0,le_1]}, T_{B(0,\frac{N}{4})} > b_1\right]$

$$\geq P_y\left[T_{B(0,\frac{N}{4})} < H_{[0,le_1]}\right] - P_y\left[T_{B(0,\frac{N}{4})} \leq b_1\right].$$

Note that the $d-1$-dimensional projection of X obtained by omitting the first coordinate is a $d-1$-dimensional random walk with a geometric delay of constant parameter. Hence, one finds that for $y \in \partial[0, le_1]$,

(3.29) $\quad P_y\left[T_{B(0,\frac{N}{4})} < H_{[0,le_1]}\right] \geq \frac{d-1}{d}(1 - q(d-1)),$

3. Profusion of logarithmic components until time a_1

where $q(.)$ is as below (1.1) and we have used $(d-1)/d$ to bound from below the probability that the projected random walk, if starting from 0, leaves 0 in its first step. By translation invariance, for $N \geq c$, the second probability on the the right-hand side of (3.28) is bounded from above by $P_0\bigl[T_{B(0,\frac{N}{8})} \leq b_1\bigr] \leq \exp\bigl\{-cN^{\frac{1}{3}-\frac{1}{200}}\bigr\}$, with (2.4), where $a = \sqrt{b_1}$ and $b = \bigl[\frac{N}{8}\bigr]$, cf. (2.5). Hence, we find that

(3.30) $$L_3 \geq c.$$

Inserting the lower bounds on L_1, L_2 and L_3 from (3.26), (3.27) and (3.30) into (3.24) and then using (3.23), we have shown (3.20) and therefore completed the proof of Lemma 3.3. □

PROOF OF LEMMA 3.4. We denote the events on the left-hand side of (3.7) by A and B, i.e.

$$A = \{|\mathcal{I}_E| \geq [N^\nu]\}, \quad B = \bigcap_{n=0}^{a_1-a_0} \{X_{[0,n]} \cap X_{[n+a_0,a_1]} = \emptyset\}.$$

We need to show that, if $A\cap B$ occurs, then we can find $[N^\nu]$ segments of length l as components of the vacant set left until time a_1. Informally, the reasoning goes as follows: for any of the $[N^\nu]$ events $\mathcal{A}_{i,E}$ occurring on A, cf. (3.4), the random walk produces in the time interval $(i-1)b_1 + [0, b_1/2]$ a component of the vacant set consisting of a segment of length l and this segment remains unvisited for a further duration of $[b_1/2]$, much larger than a_0, cf. (2.6). However, when B occurs, after a time interval of length a_0 has elapsed, the random walk does not revisit any point on the visited boundary of the segment appearing in any of the occurring events $\mathcal{A}_{i,E}$. It follows that the segments appearing in the $[N^\nu]$ different occurring events $\mathcal{A}_{i,E}$ are distinct, unvisited and have a completely visited boundary. More precisely, we fix any $N \geq c$ such that

(3.31) $$a_0 \leq \frac{b_1}{2},$$

and assume that the events A and B both occur. We pick $1 \leq i_1 < i_2 < \ldots < i_{[N^\nu]} \leq [a_1/b_1]$ such that the events $\mathcal{A}_{i_j,E}$ occur, and denote one of the segments of the form $[x, x+le_1]$ appearing in the definition of $\mathcal{A}_{i_j,E}$ by S_j, cf. (3.4). The proof will be complete once we have shown that

$$X_{[0,a_1]} \supseteq \partial S_j,\ X_{[0,a_1]} \cap S_j = \emptyset \text{ and } S_j \neq S_{j'},$$
$$\text{for any } j, j' \in \{1,\ldots,[N^\nu]\},\ j < j'.$$

The fact $X_{[0,a_1]} \supseteq \partial S_j$ follows directly from the occurrence of the event $\mathcal{A}_{i_j,E}$ on A, cf. (3.4). To see that $X_{[0,a_1]} \cap S_j = \emptyset$, note first that by definition of $\mathcal{A}_{i_j,E}$,

(3.32) $$X_{[0,i_j b_1]} \cap S_j = \emptyset.$$

In particular, this implies that $X_{[i_jb_1,a_1]} \not\subseteq S_j$ and that for any $x \in S_j$, there is a point $x' \in \partial S_j$ such that $d\left(x, X_{[i_jb_1,a_1]}\right) \geq d\left(x', X_{[i_jb_1,a_1]}\right)$, hence

(3.33) $$d\left(S_j, X_{[i_jb_1,a_1]}\right) \geq d\left(\partial S_j, X_{[i_jb_1,a_1]}\right).$$

Moreover, one has on $\mathcal{A}_{i_j,E}$ that $\partial S_j \subseteq X_{[0,i_jb_1-b_1/2]}$, and by (3.31),

$$X_{[0,i_jb_1-b_1/2]} \subseteq X_{[0,i_jb_1-a_0]}.$$

Since B occurs, this yields

(3.34) $$\partial S_j \cap X_{[i_jb_1,a_1]} = \emptyset,$$

and hence by (3.33), $S_j \cap X_{[i_jb_1,a_1]} = \emptyset$. With (3.32) we deduce that $X_{[0,a_1]} \cap S_j = \emptyset$, as required. Finally, we need to show that $S_j \neq S_{j'}$ for $j < j'$. To this end, note that on $\mathcal{A}_{i_{j'},E}$, $X_{[i_jb_1,a_1]} \supseteq X_{[(i_{j'}-1)b_1,a_1]} \supseteq \partial S_{j'}$, and hence

$$d\left(\partial S_j, \partial S_{j'}\right) \geq d\left(\partial S_j, X_{[i_jb_1,a_1]}\right) \overset{(3.34)}{>} 0.$$

Hence (3.7) is proved and the proof of Lemma 3.4 is complete. □

The statement (3.2) is now a direct consequence of (3.3), (3.6) and (3.7), so that the proof of Proposition 3.1 is finished. □

4. Survival of a logarithmic segment

This section is devoted to the preparation of the second part of the proof of Theorem 1.1, that is claim (1.6). We show that at least one of the $[N^\nu]$ isolated segments produced until time a_1 remains unvisited by the walk until time uN^d. As mentioned in the introduction, the strategy is to use a lower bound of e^{-cul} on the probability that one fixed segment remains unvisited until a (random) time larger than uN^d. The desired statement (1.6) would then be an easy consequence if the events $\{X_{[0,uN^d]} \cap [x, x+le_1] = \emptyset\}$ were independent for different $x \in E$, but this is obviously not the case. However, a technique developed in [6] allows to bound the covariance between such events for sufficiently distant points x and x' and with uN^d replaced by the random time $D^x_{l^*(u)}$. Here, D^x_k is defined as the end of the k-th excursion in and out of concentric boxes of suitable size centered at $x \in E$, and $l^*(u)$ is chosen such that with high probability, $D^x_{l^*(u)} \geq uN^d$, see (4.6) and (4.7) below. The variance bounds from [6] and the above-mentioned estimates yield the desired claim in Proposition 4.1. In order to state this proposition, we introduce the integer-valued random variable $\Gamma^J_{[s,t]}$ for $s,t \geq 0$ and $J \subseteq E$, counting the number of sites x in J such that the segment $[x, x+le_1]$ is not visited by $X_{[s,t]}$, i.e.

(4.1) $$\Gamma^J_{[s,t]} = \sum_{x \in J} \mathbf{1}_{\{X_{[s,t]} \cap [x,x+le_1]=\emptyset\}}.$$

4. Survival of a logarithmic segment

The following proposition asserts that for $\nu > 0$ and an arbitrary set J of size at least $[N^\nu]$, when $u > 0$ is chosen small enough, $\Gamma^J_{[0,uN^d]}$ is not zero with P_0-probability tending to 1 as N tends to infinity. Combined with the application of the Markov property at time a_1, it will play a crucial role in the proof of (1.6), cf. (5.2) below.

PROPOSITION 4.1. ($d \geq 4, 0 < \nu < 1$) For l as in (1.3),

$$(4.2) \qquad \lim_N \inf_{\substack{J \subseteq E \\ |J| \geq [N^\nu]}} P_0 \left[\Gamma^J_{[0,uN^d]} \geq 1 \right] = 1, \text{ for small } u > 0.$$

PROOF. Throughout the proof, we say that a statement applies "for large N" if the statement applies for all N larger than a constant depending only on d and ν. The central part of the proof is an application of a technique for estimating the covariance of "local functions" of distant subsets of points in the torus, developed in [6]. In order to apply the corresponding result from [6], we set

$$(4.3) \qquad L = \left[(\log N)^2 \right]$$

and, for large N, consider any positive integer r such that

$$(4.4) \qquad 10L \leq r \leq \left[N^{\frac{\nu}{d}} \right].$$

Note that L and r then satisfy (3.1) of [6]. We then define the nested boxes

$$(4.5) \qquad C(x) = B(x, L), \text{ and } \tilde{C}(x) = B(x, r).$$

Finally, we consider the stopping times $(R^x_k, D^x_k)_{k \geq 1}$, the successive returns to $C(x)$ and departures from $\tilde{C}(x)$, defined as in [6], (4.8), by

$$(4.6) \qquad \begin{aligned} R^x_1 &= H_{C(x)}, \quad D^x_1 = T_{\tilde{C}(x)} \circ \theta_{R^x_1} + R^x_1, \text{ and for } n \geq 2, \\ R^x_n &= R^x_1 \circ \theta_{D^x_{n-1}} + D^x_{n-1}, \quad D^x_n = D^x_1 \circ \theta_{D^x_{n-1}} + D^x_{n-1}, \end{aligned}$$

so that $0 \leq R_1 < D_1 < \ldots < R_k < D_k < \ldots$, P-a.s. The following estimate from [6] on these returns and departures will be used:

LEMMA 4.2. ($d \geq 3, L = \left[(\log N)^2 \right], r \geq 10L, N \geq 10r$) There is a constant $c_2 > 0$, such that for $u > 0$, $x \in E$,

$$(4.7) \qquad P_0 \left[R^x_{l^*(u)} \leq uN^d \right] \leq cN^d e^{-c'uL^{d-2}}, \text{ with } l^*(u) = \left[c_2 u L^{d-2} \right].$$

PROOF OF LEMMA 4.2. The statement is the same as (4.9) in [6], except that we have here replaced P by P_0 and added an extra factor of N^d on the right-hand side of (4.7). It therefore suffices to note that

$$P \left[R^x_{l^*(u)} \leq uN^d \right] \geq \frac{1}{N^d} P_0 \left[R^x_{l^*(u)} \leq uN^d \right].$$

□

We now control the complement of the event in (4.2). To this end, fix any $J \subseteq E$ such that $|J| = [N^\nu]$ and note that

(4.8) $P_0 \left[\Gamma^J_{[0,uN^d]} = 0 \right]$

$\leq P_0 \left[\left\{ \Gamma^J_{[0,uN^d]} = 0 \right\} \cap \left\{ D^x_{l^*(u)} \geq uN^d \text{ for all } x \in E \right\} \right]$
$+ P_0 \left[\text{for some } x \in E, R^x_{l^*(u)} < D^x_{l^*(u)} < uN^d \right],$

$\overset{(4.7)}{\leq} P_0 \left[\tilde{\Gamma}_u = 0 \right] + N^c e^{-c'u(\log N)^{2(d-2)}}$, where

(4.9) $\tilde{\Gamma}_u = \sum_{x \in J} \mathbf{1}_{\{H_{[x,x+le_1]} > D^x_{l^*(u)}\}} \overset{(\text{def.})}{=} \sum_{x \in J} h(x),$

and $l^*(u)$ was defined in (4.7). In order to bound the probability in (4.8), we need an estimate on the variance of $\tilde{\Gamma}_u$. This estimate can be obtained by using the bound on the covariance of $h(x)$ and $h(y)$ for x and y sufficiently far apart, derived in [6]. To this end, one first notes that

$$\text{var}_{P_0}\left(\tilde{\Gamma}_u\right) = \text{var}_{P_0}\left(\sum_{x \in J} h(x)\right)$$
$$\leq c\left(N^\nu r^d + r^{2d}\right) + N^{2\nu} \sup_{\substack{x,y \in E \\ |x-y|_\infty \geq 2r+3 \\ x,y \notin \tilde{C}(0)}} \text{cov}_{P_0}\left(h(x), h(y)\right).$$

In the proof of Proposition 4.2 in [6], the covariance in the last supremum is bounded from above by $cu\frac{L^d}{r}$ (cf. [6], above (4.44)). Since $r^d \leq N^\nu$ (cf. (4.4)), we therefore have

(4.10) $\text{var}_{P_0}\left(\tilde{\Gamma}_u\right) \leq c\left(r^d N^\nu + u\frac{N^{2\nu} L^d}{r}\right).$

Below, we will show that

(4.11) $P_0\left[H_{[x,x+le_1]} > D^x_{l^*(u)}\right] \geq ce^{-c'ul}$, when $0 \notin [x, x+le_1]$.

Before we prove this claim, we show how to deduce Proposition 4.1 from the above. It follows from (4.11) that for large N,

$$E_0\left[\tilde{\Gamma}_u\right] = \sum_{x \in J} P_0\left[H_{[x,x+le_1]} > D^x_{l^*(u)}\right] \geq c_3 N^\nu e^{-c_4 ul}.$$

Hence for large N, one has

(4.12) $P_0\left[\tilde{\Gamma}_u = 0\right] \leq P_0\left[\tilde{\Gamma}_u < E_0\left[\tilde{\Gamma}_u\right] - \frac{c_3}{2} N^\nu e^{-c_4 ul}\right]$

$\leq c \, \text{var}_{P_0}(\tilde{\Gamma}_u) N^{-2\nu} e^{cul} \overset{(4.10)}{\leq} c\left(\frac{r^d}{N^\nu} + u\frac{L^d}{r}\right) e^{cul}.$

4. Survival of a logarithmic segment

We now choose $r = \left[(L^d N^\nu)^{\frac{1}{d+1}}\right]$, so that with (4.3) one has

$$cr \leq (\log N)^{\frac{2d}{d+1}} N^{\frac{\nu}{d+1}} \leq c'r$$

and r satisfies (4.4) for large N. Inserting these choices of r, L and l from (1.3) into the estimate (4.12), one obtains

$$P_0\left[\tilde{\Gamma}_u = 0\right] \leq c(1+u)(\log N)^c N^{-\frac{\nu}{d+1}+cu}.$$

For $u > 0$ chosen sufficiently small, the right-hand side tends to 0 as $N \to \infty$. With (4.8) and monotonicity of Γ^J in J, this proves (4.2). There only remains to show (4.11).

First, the strong Markov property applied at time $T_{C(x)}$ yields that

$$P_0\left[H_{[x,x+le_1]} > D^x_{l^*(u)}\right] \geq P_0\left[H_{[x,x+le_1]} > D^x_{l^*(u)}, T_{C(x)} < H_{[x,x+le_1]}\right]$$
$$\geq P_0\left[T_{C(x)} < H_{[x,x+le_1]}\right] \inf_{y \notin C(x)} P_y\left[H_{[x,x+le_1]} > D^x_{l^*(u)}\right].$$

For x such that $0 \notin [x, x+le_1]$, transience of simple random walk in dimension $d-1$ implies that $P_0\left[T_{C(x)} < H_{[x,x+le_1]}\right] \geq c > 0$, and hence,

(4.13) $\quad P_0\left[H_{[x,x+le_1]} > D^x_{l^*(u)}\right] \geq c \inf_{y \notin C(x)} P_y\left[H_{[x,x+le_1]} > D^x_{l^*(u)}\right].$

The application of the strong Markov property at the times

$$R^x_{l^*(u)}, R^x_{l^*(u)-1}, \ldots, R^x_1$$

then yields

(4.14) $\quad \inf_{y \notin C(x)} P_y\left[H_{[x,x+le_1]} > D^x_{l^*(u)}\right]$
$$\geq \inf_{y \in \partial(C(x)^c)} P_y\left[H_{[x,x+le_1]} > D^x_1\right]^{l^*(u)}.$$

From the right-hand estimate of (2.2) on the hitting probability with $A = [x, x+le_1]$ and $B = \tilde{C}(x)$ and the trivial lower bound of 1 for the denominator of the right-hand side, one obtains that

$$\sup_{y \in \partial(C(x)^c)} P_y\left[H_{[x,x+le_1]} \leq D^x_1\right] \leq \sup_{y \in \partial(C(x)^c)} \sum_{z \in [x,x+le_1]} g^{\tilde{C}(x)}(y,z) \leq cl L^{-(d-2)},$$

with the Green function estimate from [19], Theorem 1.5.4 in the last step. Inserting this bound into (4.14), one deduces that

$$\inf_{y \notin C(x)} P_y\left[H_{[x,x+le_1]} > D^x_{l^*(u)}\right] \geq \left(1 - cl L^{-(d-2)}\right)^{l^*(u)}$$
$$\geq e^{-c'lL^{-(d-2)}l^*(u)} \geq e^{-c''ul}.$$

With (4.13), this shows (4.11) and thus completes the proof of Proposition 4.1. $\square$

5. Proof of the main result

Finally, we combine the results of the two previous sections to deduce Theorem 1.1 as a corollary of Propositions 3.1 and 4.1.

PROOF OF THEOREM 1.1. Note that if the giant component O has macroscopic volume, then any component consisting only of a segment of length l must be distinct from O. In other words, one has for $N \geq c$, cf. (3.1),

$$\mathcal{G}_{\beta,uN^d} \cap \bigcup_{x \in E} \{[x, x + le_1] \subseteq E \setminus (X_{[0,uN^d]} \cup O)\}$$

$$\supseteq \mathcal{G}_{\beta,uN^d} \cap \left\{\frac{|O|}{N^d} \geq \frac{1}{2}\right\} \cap \{\mathcal{J}_{uN^d} \neq \emptyset\}.$$

In view of (1.2), it hence suffices to show that

(5.1) $\qquad \lim_N P[\mathcal{J}_{uN^d} \neq \emptyset] = 1$, for small $u > 0$.

However, the event in (5.1) occurs as soon as there are at least $[N^\nu]$, $\nu > 0$, segments of length l as components in the vacant set at time a_1, at least one of which is not visited by the random walk until time uN^d. For any $\nu > 0$ and large N (depending on ν), the probability in (5.1) is therefore bounded from below by (cf. (4.1))

$$P\left[\{|\mathcal{J}_{a_1}| \geq [N^\nu]\} \cap \left\{\Gamma^{\mathcal{J}_{a_1}}_{[a_1,uN^d]} \geq 1\right\}\right]$$

$$= \sum_{\substack{J \subseteq E \\ |J| \geq [N^\nu]}} P\left[\{\mathcal{J}_{a_1} = J\} \cap \left\{\Gamma^{J}_{[a_1,uN^d]} \geq 1\right\}\right].$$

By the simple Markov property applied at time a_1 and translation invariance, one deduces that

(5.2) $\quad P[\mathcal{J}_{uN^d} \neq \emptyset] \geq \sum_{\substack{J \subseteq E \\ |J| \geq [N^\nu]}} P[\mathcal{J}_{a_1} = J] \inf_{\substack{J' \subseteq E \\ |J'| \geq [N^\nu]}} P_0\left[\Gamma^{J'}_{[0,uN^d]} \geq 1\right]$

$$= P[|\mathcal{J}_{a_1}| \geq [N^\nu]] \inf_{\substack{J \subseteq E \\ |J| \geq [N^\nu]}} P_0\left[\Gamma^{J}_{[0,uN^d]} \geq 1\right].$$

For small $\nu > 0$, this last quantity tends to 1 as $N \to \infty$ if $u > 0$ is chosen small enough, by (3.2) and (4.2). This completes the proof of (5.1) and hence of Theorem 1.1. $\square$

REMARK 5.1.
1) With only minor modifications, the proof presented in this work shows that for $u > 0$ chosen sufficiently small, on an event of probability tending to 1 as N tends to infinity, the vacant set left until time $[uN^d]$ contains at least $[N^{c(u)}]$ segments of length l, for a constant $c(u)$ depending on d and u. Indeed, the proof of Proposition 4.1, with obvious changes, shows that for an arbitrary set $J \subseteq E$ of size at least

5. Proof of the main result

$[N^\nu]$, one has $\Gamma^J_{[0,uN^d]} \geq \frac{c_3}{2} N^\nu e^{-c_4 u l}$ with probability tending to 1 as N tends to infinity, if $u > 0$ is chosen sufficiently small, and this result can be used in the above proof to show the claim just made.

2) From results of [6] and the present work, it follows that uniqueness of a connected component of $E \setminus X_{[0,uN^d]}$ containing segments of length $[c \log N]$ holds for a certain $c = c_0$ (cf. (0.7) in [6]) and fails for a certain $c = c_1$ with overwhelming probability, when $u > 0$ is chosen sufficiently small. It is thus natural to consider the value

$$c_* = \inf\{c > 0 : \text{for small } u > 0, \lim_N P[\mathcal{O}_{c,u}] = 1\}, \text{ where}$$

$$\mathcal{O}_{c,u} \stackrel{(\text{def.})}{=} \{E \setminus X_{[0,uN^d]} \text{ contains exactly one connected component}$$
$$\text{containing segments of length } [c \log N]\}.$$

The results in [6] show in particular that $c_* < \infty$, and the present work shows that $c_* > 0$, hence c_* is non-degenerate for $d \geq d_0$. One may then ask if it is true that for arbitrary $0 < c < c_* < c'$, $\lim_N P[\mathcal{O}_{c,u}] = 0$ and $\lim_N P[\mathcal{O}_{c',u}] = 1$, when $u > 0$ is chosen sufficiently small. In fact, using results from [6], one easily deals with the case $c' > c_*$. Indeed, on the event $\mathcal{V}_{c',1/2,uN^d}$ (defined in (3.15)), the events $\mathcal{O}_{c'',u}$ increase in $c'' \leq c'$, so that one has $\mathcal{V}_{c',1/2,uN^d} \cap \mathcal{O}_{c'',u} \subseteq \mathcal{O}_{c',u}$ for $c'' \leq c'$. Since $\lim_N P[\mathcal{V}_{c',1/2,uN^d}] = 1$ for $u > 0$ chosen small enough (cf. (1.26) in [6]), this implies that if $\lim_N P[\mathcal{O}_{c'',u}] = 1$, then $\lim_N P[\mathcal{O}_{c',u}] = 1$ for any $c' > c''$. As far as the value or the large-d-behavior of c_* is concerned, only little follows from [6] and this work. While the upper bound from [6] (cf. (2.47) in [6]) behaves like $d(\log d)^{-1}$ for large d, our lower bound behaves like $(d \log d)^{-1}$ (cf. (1.3)), which leaves much scope for improvement.

3) This work shows a lower bound on non-giant components of the vacant set. Apart from the fact that vacant segments outside the giant component cannot be longer than $[c_0 \log N]$, little is known about upper bounds on such components. Although (1.2) does imply that the volume of a non-giant component of the vacant set is with overwhelming probability not larger than $(1-\gamma)N^d$ for arbitrary $\gamma \in (0,1)$, when $u > 0$ is small enough, simulations indicate that the volume of such components is typically much smaller. Further related open questions are raised in [6].

CHAPTER 3

Random walk on a discrete torus and random interlacements

We investigate the relation between the local picture left by the trajectory of a simple random walk on the torus $(\mathbb{Z}/N\mathbb{Z})^d$, $d \geq 3$, until uN^d time steps, $u > 0$, and the model of random interlacements recently introduced by Sznitman [39]. In particular, we show that for large N, the joint distribution of the local pictures in the neighborhoods of finitely many distant points left by the walk up to time uN^d converges to independent copies of the random interlacement at level u.

1. Introduction

The object of a recent article by Benjamini and Sznitman [6] was to investigate the vacant set left by a simple random walk on the $d \geq 3$-dimensional discrete torus of large side-length N up to times of order N^d. The aim of the present work is to study the connections between the microscopic structure of this set and the model of random interlacements introduced by Sznitman in [39]. Similar questions have also recently been considered in the context of random walk on a discrete cylinder with a large base, see [36].

In the terminology of [39], the interlacement at level $u \geq 0$ is the trace left on $\mathbb{Z}^d$ by a cloud of paths constituting a Poisson point process on the space of doubly infinite trajectories modulo time-shift, tending to infinity at positive and negative infinite times. The parameter u is a multiplicative factor of the intensity measure of this point process. The interlacement at level u is an infinite connected random subset of $\mathbb{Z}^d$, ergodic under translation. Its complement is the so-called vacant set at level u. In this work, we consider the distribution of the local pictures of the trajectory of the random walk on $(\mathbb{Z}/N\mathbb{Z})^d$ running up to time uN^d in the neighborhood of finitely many points with diverging mutual distance as N tends to infinity. We show that the distribution of these sets converges to the distribution of independent random interlacements at level u.

In order to give the precise statement, we introduce some notation. For $N \geq 1$, we consider the integer torus

(1.1) $$\mathbb{T} = (\mathbb{Z}/N\mathbb{Z})^d, \quad d \geq 3.$$

90   3. Random walk on a discrete torus and random interlacements

We denote with P_x, $x \in \mathbb{T}$, or P, the canonical law on $\mathbb{T}^{\mathbb{N}}$ of simple random walk on $\mathbb{T}$ starting at x, or starting with the uniform distribution ν on $\mathbb{T}$, respectively. The corresponding expectations are denoted by E_x and E, the canonical process by $X_\cdot$. Given $x \in \mathbb{T}$, the vacant configuration left by the walk in the neighborhood of x at time $t \geq 0$ is the $\{0,1\}^{\mathbb{Z}^d}$-valued random variable

(1.2) $\omega_{x,t}(.) = 1\{X_m \neq \pi_{\mathbb{T}}(.) + x, \text{ for all } 0 \leq m \leq [t]\}$,

where $\pi_{\mathbb{T}}$ denotes the canonical projection from $\mathbb{Z}^d$ onto $\mathbb{T}$. With (2.16) of [**39**], the law $\mathbb{Q}_u$ on $\{0,1\}^{\mathbb{Z}^d}$ of the indicator function of the vacant set at level $u \geq 0$ is characterized by the property

(1.3) $\mathbb{Q}_u[\omega(x) = 1, \text{ for all } x \in K] = \exp\{-u \operatorname{cap}(K)\}$,

for all finite sets $K \subseteq \mathbb{Z}^d$, where $\omega(x)$, $x \in \mathbb{Z}^d$, are the canonical coordinates on $\{0,1\}^{\mathbb{Z}^d}$, and $\operatorname{cap}(K)$ the capacity of K, see (2.14) below. In this note, we show that the joint distribution of the vacant configurations in $M \geq 1$ distinct neighborhoods of distant points $x_1, \ldots, x_M$ at time uN^d tends to the distribution of M vacant sets of independent random interlacements at level u. This result has a similar flavor to Theorem 0.1 in [**36**], which was proved in the context of random walk on a discrete cylinder.

THEOREM 1.1. ($u > 0$, $d \geq 3$) Consider $M \geq 1$ and for each $N \geq 1$, $x_1, \ldots, x_M$ points in $\mathbb{T}$ such that

(1.4) $\lim_N \inf_{1 \leq i \neq j \leq M} |x_i - x_j|_\infty = \infty$. Then

(1.5) $(\omega_{x_1, uN^d}, \ldots, \omega_{x_M, uN^d})$ converges in distribution to $\mathbb{Q}_u^{\otimes M}$

under P, as $N \to \infty$.

We now make some comments on the proof of Theorem 1.1. Standard arguments show that it suffices to show convergence of probabilities of the form $P[H_B > uN^d]$ with $B = \bigcup_{i=1}^M (x_i + K_i)$ and finite subsets K_i of $\mathbb{Z}^d$, where H_B denotes the time until the first visit to the set $B \subseteq \mathbb{T}$ by the random walk. Since the size of the set B does not depend on N, it is only rarely visited by the random walk for large N. It is therefore natural to expect that H_B should be approximately exponentially distributed, see Aldous [**1**], B2, p. 24. This idea is formalized by Theorem 2.1 below, quoted from Aldous and Brown [**2**]. Assuming that the distribution of H_B is well approximated by the exponential distribution with expectation $E[H_B]$, the probability $P[H_B > uN^d]$ is approximately equal to $\exp\{-uN^d/E[H_B]\}$. In order to show that this expression tends to the desired limit, which by (1.3) and (1.5) is given by $\prod_{i=1}^M \exp\{-u \operatorname{cap}(K_i)\}$, one has to show that $N^d/E[H_B]$ tends to $\sum_{i=1}^M \operatorname{cap}(K_i)$.

1. Introduction

This task is accomplished with the help of the variational characterizations of the capacity of finite subsets of $\mathbb{Z}^d$ given by the Dirichlet and Thomson principles, see (2.16) and (2.17). These principles characterize the capacity of a finite subset A of $\mathbb{Z}^d$ as the infimum over all Dirichlet forms of functions of finite support on $\mathbb{Z}^d$ taking the value 1 on A, or as the supremum over the reciprocal of energies dissipated by unit flows from A to infinity. Aldous and Fill [3] show that very similar variational characterizations involving functions and flows on $\mathbb{T}$ hold for the quantity $N^d/E[H_A]$, see (2.11) and (2.12) below. In these two variational characterizations one optimizes the same quantities as in the Dirichlet and Thomson principles, over functions on the torus of zero mean, and over unit flows on the torus from A to the uniform distribution. In the proof, we compare these two variational problems with the corresponding Dirichlet and Thomson principles and thus show the coincidence of $\lim_N N^d/E[H_B]$ with $\sum_{i=1}^M \mathrm{cap}(K_i)$. To achieve this goal, we construct a nearly optimal test function and a nearly optimal test flow for the variational problems on $\mathbb{T}$ using nearly optimal functions and a nearly optimal flow for the corresponding Dirichlet and Thomson principles.

In the case of the Dirichlet principle, this construction is rather simple and only involves shifting the functions on $\mathbb{Z}^d$ whose Dirichlet forms are almost $\mathrm{cap}(K_i)$ to the points x_i on the torus, adding and rescaling them. In the Thomson principle, we identify the torus with a box in $\mathbb{Z}^d$ and consider the unit flow from B to infinity on $\mathbb{Z}^d$ with dissipated energy equal to $\mathrm{cap}(B)^{-1}$. To obtain a flow on $\mathbb{T}$, we first restrict the flow to the box. The resulting flow then leaves charges at the boundary. In order to obtain a nearly optimal flow from B to the uniform distribution for the variational problem (2.12) on the torus, these charges need to be redirected such that they become uniformly distributed on $\mathbb{T}$, with the help of an additional flow of small energy.

The article is organized as follows:

In section 2, we state the preliminary result on the approximate exponentiality of the distribution of H_B and introduce the variational characterizations required. In section 3, we prove Theorem 1.1.

Finally, we use the following convention concerning constants: Throughout the text, c or c' denote positive constants which only depend on the dimension d, with values changing from place to place. Dependence of constants on additional parameters appears in the notation. For example, $c(M)$ denotes a constant depending only on d and M.

Acknowledgments. The author is grateful to Alain-Sol Sznitman for proposing the problem and for helpful advice.

2. Preliminaries

In this section, we introduce some notation and results required for the proof of Theorem 1.1. We denote the l_1 and l_∞-distances on $\mathbb{T}$ or $\mathbb{Z}^d$ by $|.|_1$ and $|.|_\infty$. For any points x, x' in $\mathbb{T}$ or $\mathbb{Z}^d$, we write $x \sim x'$ if x and x' are neighbors with respect to the natural graph structure, i.e. if $|x - x'|_1 = 1$. For subsets A and B of $\mathbb{T}$ or $\mathbb{Z}^d$, we write $d(A, B)$ for their mutual distance induced by $|.|_\infty$, i.e. $d(A, B) = \inf\{|x - x'|_\infty : x \in A, x' \in B\}$, $\mathrm{int} A = \{x \in A : x' \in A \text{ for all } x' \sim x\}$, as well as $\partial_{int} A$ for the interior boundary, i.e. $\partial_{int} A = A \setminus \mathrm{int} A$, and $|A|$ for the number of points in A.

We obtain a continuous-time random walk $(X_{\eta_t})_{t \geq 0}$ by defining the Poisson process $(\eta_t)_{t \geq 0}$ of parameter 1, independent of X. We write $P^{\mathbb{Z}^d}$ for the law of the simple random walk on $\mathbb{Z}^d$ and also denote the corresponding canonical process on $\mathbb{Z}^d$ as $X.$, which should not cause any confusion. For $t \geq 0$, the set of points visited by the random walk until time $[t]$ is denoted by $X_{[0,t]}$, i.e. $X_{[0,t]} = \{X_0, X_1, \ldots, X_{[t]}\}$. For any subset A of $\mathbb{T}$ or of $\mathbb{Z}^d$, we define the discrete- and continuous-time entrance times H_A and $\bar{H}_A$ as

(2.1) $\quad H_A = \inf\{n \geq 0 : X_n \in A\} \quad \text{and} \quad \bar{H}_A = \inf\{t \geq 0 : X_{\eta_t} \in A\}$,

as well as the hitting time

(2.2) $\quad\quad\quad\quad \tilde{H}_A = \inf\{n \geq 1 : X_n \in A\}$.

Note that by independence of X and η, one then has

(2.3) $\quad E[\bar{H}_A] = \sum_{n=0}^{\infty} P[H_A = n] E[\inf\{t \geq 0 : \eta_t = n\}]$

$= \sum_{n=0}^{\infty} P[H_A = n] n = E[H_A]$.

The Green function of the simple random walk on $\mathbb{Z}^d$ is defined as

(2.4) $\quad g(x, x') = E_x^{\mathbb{Z}^d}\left[\sum_{n=0}^{\infty} 1\{X_n = x'\}\right], \quad \text{for } x, x' \in \mathbb{Z}^d$.

In order to motivate the remaining definitions given in this section, we quote a result from Aldous and Brown [2], which estimates the difference between the distribution of $\bar{H}_A$ and the exponential distribution. The following theorem appears as Theorem 1 in [2] for general irreducible, finite-state reversible continuous-time Markov chains and is stated here for the continuous-time random walk $(X_{\eta_t})_{t \geq 0}$ on $\mathbb{T}$, cf. the remark after the statement.

THEOREM 2.1. ($d \geq 1$) *For any subset A of $\mathbb{T}$ and $t \geq 0$,*

(2.5) $\quad\quad \left| P[\bar{H}_A > t E[H_A]] - \exp\{-t\} \right| \leq cN^2 / E[H_A]$.

2. Preliminaries

REMARK 2.2. In (2.5), we have used (2.3) to replace $E[\bar{H}_A]$ by $E[H_A]$, as well as the fact that the spectral gap of the transition matrix of the random walk X on $\mathbb{T}$ is bounded from below by cN^{-2}. One of the many ways to show this last claim is to first find (by an explicit calculation of the eigenvalues, see, for example, [**3**], Chapter 5, Example 7) that in dimension $d = 1$, the spectral gap is given by $\rho_1 = 1 - \cos(2\pi/N) \geq cN^{-2}$. The d-dimensional random walk X on $\mathbb{T}$ can be viewed as a d-fold product chain, from which it follows that its spectral gap is equal to $\rho_1/d \geq cN^{-2}$, cf. [**28**], Lemma 2.2.11.

The main aim in the proof of Theorem 1.1 will be to obtain the limit as N tends to infinity of probabilities of the form $P[\bar{H}_A > uN^d]$. In view of (2.5), it is thus helpful to understand the asymptotic behavior of expected entrance times. To this end, we will use variational characterizations of expected entrance times involving Dirichlet forms and flows, which we now define. For a real-valued function f on $E = \mathbb{T}$ or $\mathbb{Z}^d$, we define the Dirichlet form $\mathcal{E}_E$ as

$$(2.6) \qquad \mathcal{E}_E(f,f) = \frac{1}{2} \sum_{x \in E} \sum_{x' \sim x} (f(x) - f(x'))^2 \frac{1}{2d}.$$

We write C_c for the set of real-valued functions on $\mathbb{Z}^d$ of finite support and denote the supremum norm of any function f by $|f|_\infty$. The integral of a function f on $\mathbb{T}$ with respect to the uniform distribution ν is denoted by $\nu(f)$ (i.e. $\nu(f) = N^{-d} \sum_{x \in \mathbb{T}} f(x)$). A flow $I = (I_{x,x'})$ on the edges of $E = \mathbb{T}$ or $\mathbb{Z}^d$ is a real-valued function on E^2 satisfying

$$(2.7) \qquad I_{x,x'} = \begin{cases} -I_{x',x} & \text{if } x \sim x', \\ 0 & \text{otherwise.} \end{cases}$$

Given a flow I, we write $|I|_\infty = \sup_{x,x' \in E} |I_{x,x'}|$ and define its dissipated energy as

$$(2.8) \qquad (I,I)_E = \frac{1}{2} \sum_{x \in E} \sum_{x' \in E} I_{x,x'}^2 \, 2d.$$

The set of all flows on the edges of E with finite energy is denoted by $F(E)$. For a flow $I \in F(E)$, the divergence div$I$ on E associates to every point in E the net flow out of it,

$$(2.9) \qquad \mathrm{div} I(x) = \sum_{x' \sim x} I_{x,x'}, \quad x \in E.$$

The net flow out of a finite subset $A \subseteq E$ is denoted by

$$(2.10) \qquad I(A) = \sum_{x \in A} \sum_{x' \sim x} I_{x,x'} = \sum_{x \in A} \mathrm{div} I(x).$$

From Aldous and Fill, Chapter 3, Proposition 41, it is known that $N^d/E[H_A]$ is given by the infimum over all Dirichlet forms of functions on $\mathbb{T}$ of zero mean and equal to 1 on A, and by the supremum over the reciprocals of energies dissipated by unit flows from A to the uniform distribution ν:

(2.11) $\quad N^d/E[H_A] = \inf\{\mathcal{E}_\mathbb{T}(f,f) : f = 1 \text{ on } A, \nu(f) = 0\}$

(2.12) $\quad\quad\quad\quad\quad = \sup\{1/(I,I)_\mathbb{T} : I \in F(\mathbb{T}), I(A) = 1 - |A|N^{-d},$
$\quad\quad\quad\quad\quad\quad \mathrm{div} I(x) = -N^{-d} \text{ for all } x \in \mathbb{T} \setminus A\}.$

These variational characterizations are very similar to the Dirichlet and Thomson principles characterizing the capacity of finite subsets of $\mathbb{Z}^d$, to which we devote the remainder of this section. A set $A \subseteq \mathbb{Z}^d$ has its associated equilibrium measure e_A on $\mathbb{Z}^d$, defined as

(2.13) $\quad e_A(x) = \begin{cases} P_x^{\mathbb{Z}^d}[\tilde{H}_A = \infty] & \text{if } x \in A, \\ 0 & \text{if } x \in \mathbb{Z}^d \setminus A. \end{cases}$

The capacity of A is defined as the total mass of e_A,

(2.14) $\quad\quad\quad\quad \mathrm{cap}(A) = e_A(\mathbb{Z}^d).$

For later use, we record that the following expression for the hitting probability of A is obtained by conditioning on the time and location of the last visit to A and applying the simple Markov property:

(2.15) $\quad P_x^{\mathbb{Z}^d}[H_A < \infty] = \sum_{x' \in A} g(x,x') e_A(x'), \quad \text{for } x \in \mathbb{Z}^d.$

The Dirichlet and Thomson principles assert that $\mathrm{cap}(A)$ is obtained by minimizing the Dirichlet form over all functions on $\mathbb{Z}^d$ of compact support equal to 1 on A, or by maximizing the reciprocal of the energy dissipated by so-called unit flows from A to infinity:

PROPOSITION 2.3. $(d \geq 3, A \subseteq \mathbb{Z}^d, |A| < \infty)$

(2.16) $\quad\quad \mathrm{cap}(A) = \inf\{\mathcal{E}_{\mathbb{Z}^d}(f,f) : f \in C_c, f = 1 \text{ on } A\}$

(2.17) $\quad\quad\quad\quad\quad = \sup\{1/(I,I)_{\mathbb{Z}^d} : I \in F(\mathbb{Z}^d), I(A) = 1,$
$\quad\quad\quad\quad\quad\quad \mathrm{div} I(x) = 0, \text{ for all } x \in \mathbb{Z}^d \setminus A\}.$

Moreover, the unique maximizing flow I^A in the variational problem (2.17) satisfies

(2.18) $\quad I_{x,x'}^A = -(2d \, \mathrm{cap}(A))^{-1}(P_{x'}^{\mathbb{Z}^d}[H_A < \infty] - P_x^{\mathbb{Z}^d}[H_A < \infty]),$

for all neighbors x and x' in $\mathbb{Z}^d$.

PROOF. By collapsing the set A to a point (see for example [3], Chapter 2, Section 7.3), it suffices to consider a general transient graph G instead of $\mathbb{Z}^d$ and $A = \{a\}$, for a vertex a in G. The proof for this case can be found in [32]: Theorem 3.41 shows (2.16) above and Theorem 3.25 with $\iota = 1_{\{a\}}$ (in the notation of [32]; allowed by Theorem 3.30 and transience of the simple random walk in dimension $d \geq 3$), combined with Corollary 2.14, yields the above claims (2.17) and (2.18). □

3. Proof

With the results of the last section, we are now ready to give the proof of Theorem 1.1.

PROOF OF THEOREM 1.1. Take any finite subsets $K_1, \ldots K_M$ of $\mathbb{Z}^d$ and, using the notations of the theorem, set $B = \bigcup_{i=1}^M (x_i + K_i)$. Note that the collection of events $\{\omega(x) = 1 \text{ for all } x \in K\}$ as K varies over finite subsets of $\mathbb{Z}^d$ forms a π-system generating the canonical product σ-algebra on $\{0,1\}^{\mathbb{Z}^d}$. By compactness of the set of probability measures on $(\{0,1\}^{\mathbb{Z}^d})^M$, our claim will follow once we show that

$$(3.1) \qquad \lim_N P[H_B > uN^d] = \prod_{i=1}^M e^{-u \operatorname{cap}(K_i)}.$$

As we now explain, we can replace H_B by its continuous-time analog $\bar{H}_B$ in (3.1). Indeed, assume (3.1) holds with H_B replaced by $\bar{H}_B$. By the law of large numbers, one has $\eta_t/t \to 1$ a.s., as t tends to infinity (see, for example [**15**], Chapter 1, Theorem 7.3), and it then follows that, for $0 < \epsilon < u$,

$$\limsup_N P[H_B > uN^d] = \limsup_N P[X_{[0, uN^d]} \cap B = \emptyset]$$
$$\leq \limsup_N P[X_{[0, \eta_{(u-\epsilon)N^d}]} \cap B = \emptyset]$$
$$= \limsup_N P[\bar{H}_B > (u-\epsilon)N^d] = \prod_{i=1}^M e^{-(u-\epsilon) \operatorname{cap}(K_i)},$$

and similarly,

$$\liminf_N P[H_B > uN^d] \geq \liminf_N P[X_{[0, \eta_{(u+\epsilon)N^d}]} \cap B = \emptyset]$$
$$= \liminf_N P[\bar{H}_B > (u+\epsilon)N^d] = \prod_{i=1}^M e^{-(u+\epsilon) \operatorname{cap}(K_i)}.$$

Letting ϵ tend to 0 in the last two estimates, one deduces the desired result. By the above observations and (2.5) with $A = B$ and $t = uN^d/E[H_B]$, all that is left to prove is that

$$(3.2) \qquad \lim_N \frac{N^d}{E[H_B]} = \sum_{i=1}^M \operatorname{cap}(K_i).$$

The claim (3.2) will be shown by using the variational characterizations (2.11), (2.12), (2.16) and (2.17). To this end, we map the torus $\mathbb{T}$ to a subset of $\mathbb{Z}^d$ in the following way: We choose a point x_* in $\mathbb{T}$ as the origin and then define the bijection $\psi : \mathbb{T} \to \mathbb{T}' = \{0, \ldots, N-1\}^d$ such that $\pi_{\mathbb{T}}(\psi(x_* + x)) = x$ for $x \in \mathbb{T}$, where $\pi_{\mathbb{T}}$ denotes the canonical projection from $\mathbb{Z}^d$ onto $\mathbb{T}$. Since there are only M points x_i, we can

96   3. Random walk on a discrete torus and random interlacements

choose x_* in such a way that in $\mathbb{T}' \subseteq \mathbb{Z}^d$, $\psi(B)$ remains at a distance of order N from the interior boundary of $\mathbb{T}'$, i.e. such that for $N \geq c(M)$,

(3.3) $\qquad d(\psi(B), \partial_{int}\mathbb{T}') \geq c'(M)N.$

We define the subsets C and S of $\mathbb{T}$ as the preimages of $\text{int}\,\mathbb{T}'$ and $\partial_{int}\mathbb{T}'$ under ψ, i.e.

(3.4) $\qquad C = \psi^{-1}(\text{int}\,\mathbb{T}'), \quad \text{and} \quad S = \psi^{-1}(\partial_{int}\mathbb{T}').$

For $\epsilon > 0$, we now consider functions $f_i \in C_c$ (see above (2.7)) such that $f_i = 1$ on K_i and

(3.5) $\qquad \mathcal{E}_{\mathbb{Z}^d}(f_i, f_i) \leq \text{cap}(K_i) + \epsilon, \text{ for } i = 1, \ldots, M, \text{ cf. (2.16)}.$

Defining $\tau_x : \mathbb{T} \to \mathbb{Z}^d$ by $\tau_x(x') = \psi(x') - \psi(x)$ for $x, x' \in \mathbb{T}$, we construct the function f by shifting the functions f_i to the points x_i, subtracting their means and rescaling so that f equals 1 on B (for large N):

$$f = \frac{\sum_{i=1}^{M} f_i \circ \tau_{x_i} - \nu\left(\sum_{i=1}^{M} f_i \circ \tau_{x_i}\right)}{1 - \nu\left(\sum_{i=1}^{M} f_i \circ \tau_{x_i}\right)}.$$

Note that by the hypothesis (1.4) and our choice (3.3) of the origin, the finite supports of the functions $f_i(. - \psi(x_i))$ intersect neither each other nor $\partial_{int}\mathbb{T}'$ for $N \geq c(M)$. One can then easily check that for $N \geq c(M, \epsilon)$ we have $f = 1$ on B and $\nu(f) = 0$. It therefore follows from (2.11) that

$$\limsup_N N^d/E[H_B] \leq \limsup_N \mathcal{E}_{\mathbb{T}}(f, f) \overset{(f_i \in C_c, (1.4))}{=} \sum_{i=1}^{M} \mathcal{E}_{\mathbb{Z}^d}(f_i, f_i) \overset{(3.5)}{\leq} \sum_{i=1}^{M} \text{cap}(K_i) + M\epsilon.$$

Letting ϵ tend to 0, one deduces that

(3.6) $\qquad \limsup_N N^d/E[H_B] \leq \sum_{i=1}^{M} \text{cap}(K_i).$

In order to show the other half of (3.2), we proceed similarly, with the help of the variational characterizations (2.12) and (2.17). We consider the flow $I^{\psi(B)} \in F(\mathbb{Z}^d)$ such that

(3.7) $\qquad I^{\psi(B)}(\psi(B)) = 1,$

(3.8) $\qquad \text{div} I^{\psi(B)}(z) = 0 \text{ for all } z \in \mathbb{Z}^d \setminus \psi(B), \text{ and}$

(3.9) $\qquad 1/(I^{\psi(B)}, I^{\psi(B)})_{\mathbb{Z}^d} = \text{cap}(\psi(B)), \text{ cf. (2.17), (2.18)}.$

The aim is to now construct a flow of similar total energy satisfying the conditions imposed in (2.12). To this end, we first define the flow

3. Proof

$I^* \in F(\mathbb{T})$ by restricting the flow $I^{\psi(B)}$ to $\mathbb{T}'$, i.e. we set

(3.10) $$I^*_{x,x'} = I^{\psi(B)}_{\psi(x),\psi(x')} \text{ for } x, x' \in \mathbb{T}.$$

We now need a flow $J \in F(\mathbb{T})$ such that $I^* + J$ is a unit flow from B to the uniform distribution on $\mathbb{T}$. Essentially, J has to redirect some of the charges $(\text{div}I^*)1_S$ left by I^* on the set S, such that these charges become uniformly distributed on the torus, and the energy dissipated by J has to decay as N tends to infinity. The following proposition yields the required flow J:

PROPOSITION 3.1. $(d \geq 1)$ There is a flow $J \in F(\mathbb{T})$ such that

(3.11) $$\text{div}J(x) + (\text{div}I^*)1_S(x) = -N^{-d}, \text{ for any } x \in \mathbb{T}, \text{ and}$$

(3.12) $$|J|_\infty \leq c(M)N^{1-d}.$$

Before we prove Proposition 3.1, we show how it enables to complete the proof of Theorem 1.1. Let us check that for large N, the flow $I^* + J$ satisfies the hypotheses imposed in (2.12) with $A = B$. Since by (3.3), $\psi(B)$ is contained in $\text{int}\mathbb{T}'$ for $N \geq c(M)$, one has for such N,

$$(I^* + J)(B) \stackrel{(3.10)}{=} I^{\psi(B)}(\psi(B)) + J(B) \stackrel{(3.7),(3.11)}{=} 1 - |B|N^{-d}.$$

Moreover, for any $N \geq c(M)$ and $x \in \mathbb{T} \setminus B$,

$$\text{div}(I^* + J)(x) \stackrel{(3.10)}{=} (\text{div}I^{\psi(B)})1_{\text{int}\mathbb{T}'}(\psi(x)) + (\text{div}I^*)1_S(x) + \text{div}J(x)$$

$$\stackrel{(3.8),(3.11)}{=} -N^{-d}.$$

The flow $I^* + J$ is hence included in the collection on the right-hand side of (2.12) with $A = B$ and it follows with the Minkowski inequality that

(3.13) $$E[H_B]N^{-d} \leq (I^* + J, I^* + J)_{\mathbb{T}} \leq \left((I^*, I^*)^{\frac{1}{2}}_{\mathbb{T}} + (J, J)^{\frac{1}{2}}_{\mathbb{T}}\right)^2.$$

By the bound (3.12) on $|J|_\infty$, one has $(J, J)_{\mathbb{T}} \leq c(M)(N^{1-d})^2 N^d = c(M)N^{2-d}$. Inserting this estimate together with

$$(I^*, I^*)_{\mathbb{T}} \stackrel{(3.10)}{\leq} (I^{\psi(B)}, I^{\psi(B)})_{\mathbb{Z}^d} \stackrel{(3.9)}{=} 1/\text{cap}(\psi(B))$$

into (3.13), we deduce that

(3.14) $$E[H_B]N^{-d} \leq \left(\text{cap}(\psi(B))^{-\frac{1}{2}} + c(M)N^{-(d-2)/2}\right)^2.$$

Finally, we claim that

(3.15) $$\lim_N \text{cap}(\psi(B)) = \sum_{i=1}^{M} \text{cap}(K_i).$$

Indeed, the standard Green function estimate from [19], p. 31, (1.35) implies that for $d \geq 3$,

$$P^{\mathbb{Z}^d}_x[H_{x'} < \infty] \leq g(x, x') \leq c|x - x'|^{2-d}_\infty, \quad x, x' \in \mathbb{Z}^d,$$

and claim (3.15) follows by assumption (1.4) and the definition (2.14) of the capacity. Combining (3.14) with (3.15), one infers that for $d \geq 3$,

$$\limsup_N E[H_B]N^{-d} \leq \left(\sum_{i=1}^M \mathrm{cap}(K_i)\right)^{-1}.$$

Together with (3.6), this shows (3.2) and therefore completes the proof of Theorem 1.1. □

It only remains to prove Proposition 3.1.

PROOF OF PROPOSITION 3.1. The task is to construct a flow J distributing the charges $(\mathrm{div} I^*)1_S$ uniformly on $\mathbb{T}$, observing that we want the estimate (3.12) to hold. To this end, we begin with an estimate on the order of magnitude of $\mathrm{div} I^*(x)$, for $x \in S$ and $N \geq c(M)$, where we sum over all neighbors z of $\psi(x)$ in $\mathbb{Z}^d \setminus \mathbb{T}'$:

(3.16)
$$\left|\mathrm{div} I^*(x)\right| \stackrel{(3.10)}{=} \left|\mathrm{div} I^{\psi(B)}(\psi(x)) - \sum_z I^{\psi(B)}_{\psi(x),z}\right|$$
$$\stackrel{(3.3),(3.8)}{\leq} \sum_z |I^{\psi(B)}_{\psi(x),z}|$$
$$\stackrel{(2.18)}{\leq} c \sum_z \mathrm{cap}(\psi(B))^{-1} \left|P^{\mathbb{Z}^d}_z[H_{\psi(B)} < \infty] - P^{\mathbb{Z}^d}_{\psi(x)}[H_{\psi(B)} < \infty]\right|$$
$$\stackrel{(2.15)}{\leq} c(M)N^{1-d}, \quad \text{for } x \in S,$$

where we have also used the estimate on the Green function of [19], Theorem 1.5.4, together with (3.3), for the last line. The required redirecting flow J will be constructed as the sum of two flows, K and L, both of which satisfy the estimate (3.12). The purpose of K is to redirect the charges $(\mathrm{div} I^*)1_S$, in such a way that the magnitude of the resulting charge at any given point is then bounded by $c(M)N^{-d}$, hence decreased by a factor of N^{-1}, cf. (3.16). Then, the flow L will be used to distribute the resulting charges uniformly on $\mathbb{T}$. The existence of the flow L will be a consequence of the following lemma (recall our convention concerning constants described at the end of the introduction and that ν denotes the uniform distribution on $\mathbb{T}$, cf. above (2.7)):

LEMMA 3.2. $(d \geq 1)$ For any function $h : \mathbb{T} \to \mathbb{R}$, there is a flow $L^h \in F(\mathbb{T})$, such that

(3.17) $\quad (\mathrm{div} L^h + h)(x) = \nu(h)$, for any $x \in \mathbb{T}$, and

(3.18) $\quad |L^h|_\infty \leq cN|h|_\infty.$

3. Proof

PROOF OF LEMMA 3.2. We construct the flow L^h by induction on the dimension d, and therefore write $\mathbb{T}_d$ rather than $\mathbb{T}$ throughout this proof. Furthermore, we denote the elements of $\mathbb{T}_d$ using the coordinates of $\mathbb{T}'_d$ as $\{(i_1, \ldots, i_d) : 0 \leq i_j \leq N - 1\}$.

In order to treat the case $d = 1$, define the flow L^h by letting the charges defined by h flow from 0 to $N - 1$, such that the same charge is left at any point. Precisely, we set $L^h_{N-1,0} = 0$ and $L^h_{i,i+1} = \sum_{j=0}^{i} (\nu(h) - h(j))$ for $i = 0, \ldots, N - 2$ (the values in the opposite directions being imposed by the condition (2.7) on a flow). The flow L^h then has the required properties (3.17) and (3.18).

Assume now that $d \geq 2$ and that the statement of the lemma holds in any dimension $< d$. Applying the one-dimensional case on every fiber $\{(0, y), \ldots, (N - 1, y)\} \cong \mathbb{T}_1$, $y \in \mathbb{T}_{d-1}$, with the function h^1 defined by $h^1(., y) = h(., y)$, one obtains the flows L^y supported by the edges of
$$\{(0, y), \ldots, (N - 1, y)\},$$
such that for any $i \in \mathbb{T}_1$,

$$(3.19) \qquad (\mathrm{div} L^y + h)(i, y) = N^{-1} \sum_{j=0}^{N-1} h(j, y) \text{ and}$$

$$(3.20) \qquad |L^y|_\infty \leq cN|h|_\infty.$$

We now apply the induction hypothesis on the slices
$$\mathbb{S}_i = \{(i, y) : y \in \mathbb{T}_{d-1}\} \cong \mathbb{T}_{d-1}, \quad i \in \mathbb{T}_1,$$
with the function h^2 given by
$$h^2(i, .) = N^{-1} \sum_{j=0}^{N-1} h(j, .).$$

For any $0 \leq i \leq N - 1$, we thus obtain a flow L^i supported by the edges of $\mathbb{S}_i$, such that for any $y \in \mathbb{T}_{d-1}$,

$$(3.21) \quad \mathrm{div} L^i(i, y) + N^{-1} \sum_{j=0}^{N-1} h(j, y) = N^{-(d-1)} \sum_{y' \in \mathbb{T}_{d-1}} h^2(i, y') = \nu(h)$$

and

$$(3.22) \qquad |L^i|_\infty \leq cN|h|_\infty.$$

Then equations (3.19)-(3.22) imply that the flow $L^h = \sum_{i=0}^{N-1} L^i + \sum_{y \in \mathbb{T}_{d-1}} L^y$ has the required properties. Indeed, the flows L^y have disjoint supports, as do the flows L^i, and therefore the estimate (3.18) on $|L^h|_\infty$ follows from (3.20) and (3.22). Finally, for any $x = (i, y) \in \mathbb{T}_1 \times \mathbb{T}_{d-1} = \mathbb{T}_d$, (3.19) and (3.21) together yield
$$(\mathrm{div} L^h + h)(x) = \mathrm{div} L^i(i, y) + \mathrm{div} L^y(i, y) + h(i, y) = \nu(h),$$

hence (3.17). This concludes the proof of Lemma 3.2. □

We now complete the proof of Proposition 3.1. To this end, we construct the auxiliary flow K described above Lemma 3.2. Set $g = (\text{div} I^*)1_S$. Writing $e_1, \ldots, e_d$ for the canonical basis of $\mathbb{R}^d$, choose a mapping $e : S \to \{\pm e_1, \ldots, \pm e_d\}$ such that $\mathbb{F}'_x \stackrel{(\text{def.})}{=} \{\psi(x), \psi(x) + e(x), \ldots, \psi(x) + (N-1)e(x)\} \subseteq \mathbb{T}'$ (whenever there are more than one possible choices for $e(x)$, take one among them arbitrarily), and define the fiber $\mathbb{F}_x = \psi^{-1}(\mathbb{F}'_x)$.

Observe that any point $x \in \mathbb{T}$ only belongs to the d different fibers $x + [0, N-1]e_i$, $i = 1, \ldots, d$. Moreover, we claim that for any $\mathbb{F} \in \{\mathbb{F}_x\}_{x \in S}$, there are at most 2 points $x \in S$ such that $\mathbb{F}_x = \mathbb{F}$. Indeed, suppose that $\mathbb{F}_x = \mathbb{F}_{x'}$ for $x, x' \in S$. Then $\psi(\mathbb{F}_x) = \psi(\mathbb{F}_{x'})$ implies that $\psi(x') = \psi(x) + ke(x)$ for some $k \in \{0, \ldots, N-1\}$ and that either $e(x) = e(x')$ or $e(x) = -e(x')$. If $e(x) = e(x')$, then for $\psi(\mathbb{F}_{x'}) = \{\psi(x) + ke(x), \psi(x) + (k+1)e(x), \ldots, \psi(x) + (k+N-1)e(x)\}$ to be a subset of $\mathbb{T}'$, we require $k = 0$ (since $\psi(x) + Ne(x) \notin \mathbb{T}'$). Similarly, if $e(x) = -e(x')$ one needs $k = N-1$. Hence, x' can only be equal to either x or $x + (N-1)e(x)$. The above two observations on the fibers $\mathbb{F}_x$ together imply the crucial fact that any point in $\mathbb{T}$ belongs to a fiber $\mathbb{F}_x$ for at most $2d$ points $x \in S$.

We then define the flow K^x from x to $x+(N-1)e(x)$ distributing the charge $g(x)$ uniformly on the fiber $\mathbb{F}_x$. That is, the flow $K^x \in F(\mathbb{T})$ is supported by the edges of $\mathbb{F}_x$, and characterized by $K^x_{x+(N-1)e(x),x} = 0$, $K^x_{x+ie(x),x+(i+1)e(x)} = -g(x)(N-(i+1))/N$ for $i = 0, \ldots, N-2$. Observe that then $|K^x|_\infty \leq |g|_\infty$ and $|\text{div} K^x + g1_{\{x\}}|_\infty = |g(x)|/N \leq |g|_\infty/N$. Moreover, any point in $\mathbb{T}$ belongs to at most $2d$ fibers $\mathbb{F}_x$, hence to the support of at most $2d$ flows K^x. If we define the flow $K \in F(\mathbb{T})$ as $K = \sum_{x \in S} K^x$, then we therefore have

$$(3.23) \qquad |K|_\infty \leq c \max_{x \in S} |K^x|_\infty \leq c|g|_\infty,$$

as well as, for $x \in \mathbb{T}$,

$$(3.24) \qquad |(\text{div} K + g)(x)| \leq \sum_{x' \in S} |(\text{div} K^{x'} + g1_{\{x'\}})(x)|$$

$$\leq |\text{div} K^x + g1_{\{x\}}|_\infty 1_S(x) + \sum_{x' \neq x: x \in \mathbb{F}_{x'}} |\text{div} K^{x'}(x)|$$

$$\leq c|g|_\infty/N.$$

We claim that the flow $J = K + L^{\text{div} K + g}$ has the required properties (3.11) and (3.12). Indeed, using the fact that $\nu(\text{div} I) = 0$ for any flow

3. Proof

$I \in F(\mathbb{T})$,

$$\operatorname{div} J + g = \operatorname{div} L^{\operatorname{div} K + g} + \operatorname{div} K + g$$
$$\stackrel{(3.17)}{=} \nu(\operatorname{div} K + g) = \nu((\operatorname{div} I^*)1_S) = -\nu((\operatorname{div} I^*)1_C)$$
$$\stackrel{(3.10)}{=} -N^{-d} \sum_{z \in \operatorname{int} \mathbb{T}'} \operatorname{div} I_z^{\psi(B)} \stackrel{(3.8)}{=} -N^{-d} I^{\psi(B)}(\psi(B)) \stackrel{(3.7)}{=} -N^{-d}.$$

Finally, the estimates (3.18), (3.23) and (3.24) imply that

$$|J|_\infty \leq |K|_\infty + |L^{\operatorname{div} K + g}|_\infty \leq c|g|_\infty \stackrel{(3.16)}{\leq} c(M) N^{1-d}.$$

The proof of Proposition 3.1 is thus complete. □

CHAPTER 4

Random walks on discrete cylinders and random interlacements

Following the recent work of Sznitman [36], we investigate the microscopic picture induced by a random walk trajectory on a cylinder of the form $G_N \times \mathbb{Z}$, where G_N is a large finite connected weighted graph, and relate it to the model of random interlacements on infinite transient weighted graphs. Under suitable assumptions, the set of points not visited by the random walk until a time of order $|G_N|^2$ in a neighborhood of a point with $\mathbb{Z}$-component of order $|G_N|$ converges in distribution to the law of the vacant set of a random interlacement on a certain limit model describing the structure of the graph in the neighborhood of the point. The level of the random interlacement depends on the local time of a Brownian motion. The result also describes the limit behavior of the joint distribution of the local pictures in the neighborhood of several distant points with possibly different limit models. As examples of G_N, we treat the d-dimensional box of side length N, the Sierpinski graph of depth N and the d-ary tree of depth N, where $d \geq 2$.

1. Introduction

In recent works, Sznitman introduces the model of random interlacements on $\mathbb{Z}^{d+1}$, $d \geq 2$ (cf. [35], [30]), and in [36] explores its relation with the microscopic structure left by simple random walk on an infinite discrete cylinder $(\mathbb{Z}/N\mathbb{Z})^d \times \mathbb{Z}$ by times of order N^{2d}. The present work extends this relation to random walk on $G_N \times \mathbb{Z}$ running for a time of order $|G_N|^2$, where the bases G_N are given by finite weighted graphs satisfying suitable assumptions, as proposed by Sznitman in [36]. The limit models that appear in this relation are random interlacements on transient weighted graphs describing the structure of G_N in a microscopic neighborhood. Random interlacements on such graphs have been constructed in [40]. Among the examples of G_N to which our result applies are boxes of side-length N, discrete Sierpinski graphs of depth N and d-ary trees of depth N.

We proceed with a more precise description of the setup. A weighted graph $(\mathcal{G}, \mathcal{E}, w_{\cdot,\cdot})$ consists of a countable set $\mathcal{G}$ of vertices, a set $\mathcal{E}$ of unordered pairs of distinct vertices, called edges, and a weight $w_{\cdot,\cdot}$, which is a symmetric function associating to every ordered pair $(\mathsf{y}, \mathsf{y}')$ of vertices a non-negative number $w_{\mathsf{y},\mathsf{y}'} = w_{\mathsf{y}',\mathsf{y}}$, non-zero if and only

104   4. Random walks on cylinders and random interlacements

if $\{y,y'\} \in \mathcal{E}$. Whenever $\{y,y'\} \in \mathcal{E}$, the vertices y and y' are called neighbors. A path of length n in $\mathcal{G}$ is a sequence of vertices $(y_0,\ldots,y_n)$ such that y_{i-1} and y_i are neighbors for $1 \leq i \leq n$. The distance $d(y,y')$ between vertices y and y' is defined as the length of the shortest path starting at y and ending at y' and $B(y,r)$ denotes the closed ball centered at y of radius $r \geq 0$. We generally omit $\mathcal{E}$ and $w_{\cdot,\cdot}$ from the notation and simply refer to $\mathcal{G}$ as a weighted graph. A standing assumption is that $\mathcal{G}$ is connected. The random walk on $\mathcal{G}$ is defined as the irreducible reversible Markov chain on $\mathcal{G}$ with transition probabilities $p^{\mathcal{G}}(y,y') = w_{y,y'}/w_y$ for y and y' in $\mathcal{G}$, where $w_y = \sum_{y' \in \mathcal{G}} w_{y,y'}$. Then $w_y p^{\mathcal{G}}(y,y') = w_{y'} p^{\mathcal{G}}(y',y)$, so a reversible measure for the random walk is given by $w(\mathcal{A}) = \sum_{y \in \mathcal{A}} w_y$ for $\mathcal{A} \subseteq \mathcal{G}$. A bijection ϕ between subsets $\mathcal{B}$ and $\mathcal{B}^*$ of weighted graphs $\mathcal{G}$ and $\mathcal{G}^*$ is called an isomorphism between $\mathcal{B}$ and $\mathcal{B}^*$ if ϕ preserves the weights, i.e. if $w_{\phi(y),\phi(y')} = w_{y,y'}$ for all $y, y' \in \mathcal{B}$.

This setup allows the definition of a random walk $(X_n)_{n \geq 0}$ on the discrete cylinder

(1.1) $$G_N \times \mathbb{Z},$$

where G_N, $N \geq 1$, is a sequence of finite connected weighted graphs with weights $(w_{y,y'})_{y,y' \in G_N}$ and $G_N \times \mathbb{Z}$ is equipped with the weights

(1.2) $$w_{x,x'} = w_{y,y'} \mathbf{1}_{\{z=z'\}} + \frac{1}{2} \mathbf{1}_{\{y=y', |z-z'|=1\}},$$

for $x = (y,z)$, $x' = (y',z')$ in $G_N \times \mathbb{Z}$.

We will mainly consider situations where all edges of the graphs have equal weight $1/2$. The random walk X starts from $x \in G_N \times \mathbb{Z}$ or from the uniform distribution on $G_N \times \{0\}$ under suitable probabilities P_x and P defined in (2.3) and (2.4) below. We consider $M \geq 1$ and sequences of points $x_{m,N} = (y_{m,N}, z_{m,N})$, $1 \leq m \leq M$, in $G_N \times \mathbb{Z}$ with mutual distance tending to infinity. We assume that the neighborhoods around any vertex $y_{m,N}$ look like balls in a fixed infinite graph $\mathbb{G}_m$, in the sense that

(1.3) we choose an $r_N \to \infty$, such that there are isomorphisms $\phi_{m,N}$
from $B(y_{m,N}, r_N)$ to $B(o_m, r_N) \subset \mathbb{G}_m$ with $\phi_{m,N}(y_{m,N}) = o_m$
for all N.

The points not visited by the random walk in the neighborhood of $x_{m,N}$ until time $t \geq 0$ induce a random configuration of points in the limit model $\mathbb{G}_m \times \mathbb{Z}$, called the vacant configuration in the neighborhood of $x_{m,N}$, which is defined as the $\{0,1\}^{\mathbb{G}_m \times \mathbb{Z}}$-valued random variable

(1.4) $$\omega_t^{m,N}(\mathbf{x}) = \begin{cases} \mathbf{1}\{X_n \neq \Phi_{m,N}^{-1}(\mathbf{x}), \forall 0 \leq n \leq t\}, & \mathbf{x} \in B(o_m, r_N) \times \mathbb{Z}, \\ 0, & \text{otherwise, for } t \geq 0, \end{cases}$$

1. Introduction

where the isomorphism $\Phi_{m,N}$ is defined by $\Phi_{m,N}(y,z) = (\phi_{m,N}(y), z - z_{m,N})$ for (y,z) in $B(y_{m,N}, r_N) \times \mathbb{Z}$.

Random interlacements on $\mathbb{G}_m \times \mathbb{Z}$ enter the asymptotic behavior of the distribution of the local pictures $\omega^{m,N}$. For the construction of random interlacements on transient weighted graphs we refer to [40]. For our purpose it suffices to know that for a weighted graph $\mathbb{G}_m \times \mathbb{Z}$ with weights defined such that the random walk on it is transient, the law $\mathbb{Q}_u^{\mathbb{G}_m \times \mathbb{Z}}$ on $\{0,1\}^{\mathbb{G}_m \times \mathbb{Z}}$ of the indicator function of the vacant set of the random interlacement at level $u \geq 0$ on $\mathbb{G}_m \times \mathbb{Z}$ is characterized by, cf. equation (1.1) of [40],

$$(1.5) \quad \mathbb{Q}_u^{\mathbb{G}_m \times \mathbb{Z}}[\omega(\mathbf{x}) = 1, \text{ for all } \mathbf{x} \in \mathbb{V}] = \exp\{-u \operatorname{cap}^m(\mathbb{V})\},$$
for all finite subsets $\mathbb{V}$ of $\mathbb{G}_m \times \mathbb{Z}$,

where $\omega(\mathbf{x})$, $\mathbf{x} \in \mathbb{G}_m \times \mathbb{Z}$, are the canonical coordinates on $\{0,1\}^{\mathbb{G}_m \times \mathbb{Z}}$, and $\operatorname{cap}^m(\mathbb{V})$ the capacity of $\mathbb{V}$ as defined in (2.7) below.

The main result of the present work requires the assumptions A1-A10 on the graph G_N, which we discuss below. In order to state the result, we have yet to introduce the local time of the $\mathbb{Z}$-projection $\pi_\mathbb{Z}(X)$ of X, defined as

$$(1.6) \quad L_n^z = \sum_{l=0}^{n-1} \mathbf{1}_{\{\pi_\mathbb{Z}(X_l) = z\}}, \text{ for } z \in \mathbb{Z}, n \geq 1,$$

as well as the canonical Wiener measure W and a jointly continuous version $L(v,t)$, $v \in \mathbb{R}$, $t \geq 0$, of the local time of the canonical Brownian motion. The main result asserts that under suitable hypotheses the joint distribution of the vacant configurations in the neighborhoods of $x_{1,N}, \ldots, x_{M,N}$ and the scaled local times of the $\mathbb{Z}$-projections of these points at a time of order $|G_N|^2$ converges as N tends to infinity to the joint distribution of the vacant sets of random interlacements on $\mathbb{G}_m \times \mathbb{Z}$ and local times of a Brownian motion. The levels of the random interlacements depend on the local times, and conditionally on the local times, the random interlacements are independent. Here is the precise statement:

THEOREM 1.1. *Assume A1-A10 (see below (2.9)), as well as*

$$(1.7) \quad \frac{w(G_N)}{|G_N|} \xrightarrow{N \to \infty} \beta, \text{ for some } \beta > 0,$$

and for all $1 \leq m \leq M$,

$$\frac{z_{m,N}}{|G_N|} \xrightarrow{N \to \infty} v_m, \text{ for some } v_m \in \mathbb{R},$$

which is in fact assumption A4, see below. Then the graphs $\mathbb{G}_m \times \mathbb{Z}$ are transient and as N tends to infinity, the $\prod_{m=1}^{M} \{0,1\}^{\mathbb{G}_m} \times \mathbb{R}_+^M$-valued

random variables
$$\left(\omega^{1,N}_{\alpha|G_N|^2}, \ldots, \omega^{M,N}_{\alpha|G_N|^2}, \frac{L^{z_{1,N}}_{\alpha|G_N|^2}}{|G_N|}, \ldots, \frac{L^{z_{M,N}}_{\alpha|G_N|^2}}{|G_N|}\right), \quad \alpha > 0, \ N \geq 1,$$
defined by (1.4) and (1.6), with r_N *and* $\phi_{m,N}$ *chosen in (5.1) and (5.2), converge in joint distribution under* P *to the law of the random vector*
$$(\omega_1, \ldots, \omega_M, U_1, \ldots, U_M)$$
with the following distribution: the variables $(U_m)_{m=1}^M$ *are distributed as*
$$((1+\beta)L(v_m, \alpha/(1+\beta)))_{m=1}^M$$
under W, *and conditionally on* $(U_m)_{m=1}^M$, *the variables* $(\omega_m)_{m=1}^M$ *have joint distribution*
$$\prod_{1 \leq m \leq M} \mathbb{Q}^{\mathbb{G}_m \times \mathbb{Z}}_{U_m/(1+\beta)}.$$

REMARK 1.2. Sznitman proves a result analogous to Theorem 1.1 in [36], Theorem 0.1, for G_N given by $(\mathbb{Z}/N\mathbb{Z})^d$ and $\mathbb{G}_m = \mathbb{Z}^d$ for $1 \leq m \leq M$. This result is covered by Theorem 1.1 by choosing, for any y and y' in $(\mathbb{Z}/N\mathbb{Z})^d$, $w_{y,y'} = 1/2$ if y and y' are at Euclidean distance 1 and $w_{y,y'} = 0$ otherwise. Then the random walk X on $(\mathbb{Z}/N\mathbb{Z})^d \times \mathbb{Z}$ with weights as in (1.2) is precisely the simple random walk considered in [36]. We then have $\beta = d$ in (1.7) and recover the result of [36], noting that the factor $1/(1+d)$ appearing in the law of the vacant set cancels with the factor $w_{\mathsf{x}} = d+1$ in our definition of the capacity (cf. (2.7)), different from the one used in [36] (cf. (1.7) in [36]).

We now make some comments on the proof of Theorem 1.1. In order to extract the relevant information from the behavior of the $\mathbb{Z}$-component of the random walk, we follow the strategy in [36] and use a suitable version of the partially inhomogeneous grids on $\mathbb{Z}$ introduced there. Results from [36] show that the total time elapsed and the scaled local time of a simple random walk on $\mathbb{Z}$ can be approximated by the random walk restricted to certain stopping times related to these grids. The difficulty that arises in the application of these results in our setup is that unlike in [36], the $\mathbb{Z}$-projection of our random walk X is not a Markov process. Indeed, the $\mathbb{Z}$-projection is delayed at each step for an amount of time that depends on the current position of the G_N-component. In order to overcome this difficulty, we decouple the $\mathbb{Z}$-component of the random walk from the G_N-component by introducing a continuous-time process $\mathsf{X} = (\mathsf{Y}, \mathsf{Z})$, such that the G_N- and $\mathbb{Z}$-components Y and Z are independent and such that the discrete skeleton of X is the random walk X on $G_N \times \mathbb{Z}$. It is not trivial to regain information about the random walk X after having switched to continuous time, because the waiting times of the process X depend on the

1. Introduction

steps of the discrete skeleton X and are in particular not iid. We therefore prove in Theorem 5.1 the continuous-time version of Theorem 1.1 first, essentially by using an abstraction of the arguments in [**36**] and making frequent use of the independence of the G_N- and $\mathbb{Z}$-components of X, and defer the task of transferring the result to discrete time to later.

Let us make a few more comments on the partially inhomogeneous grids just mentioned. Every point of these grids is a center of two concentric intervals $I \subset \tilde{I}$ with diameters of order d_N and $h_N \gg d_N$, where h_N is also the order of the mesh size of the grids throughout $\mathbb{Z}$. The definition of the grids ensures that all points $z_{m,N}$ are covered by the smaller intervals, hence the partial inhomogeneity. We then consider the successive returns to the intervals I and departures from $\tilde{I}$ of the discrete skeleton Z of Z. According to a result from [**36**] (see Proposition 3.3 below) and Lemma 3.4, these excursions contain all the relevant information needed to approximate the total time elapsed and to relate the scaled local time $\mathsf{L}^{z_{m,N}}_{\alpha|G_N|^2}/|G_N|$ of Z (see (2.6)) to the number of returns of Z to the box containing $z_{m,N}$. For these estimates to apply, the mesh size h_N of the grids has to be smaller than the square root of the total number of steps of the walk, i.e. less than $|G_N|$. At the same time, we shall need h_N to be larger than the square root of the relaxation time λ_N^{-1} of G_N, so that the G_N-component Y approaches its stationary, i.e. uniform, distribution between different excursions. This motivates the condition A2, see below (2.9), on the spectral gap λ_N of G_N.

Once the partially inhomogeneous grids are introduced, the law $\mathbb{Q}^{\mathbb{G}_m \times \mathbb{Z}}$ of the vacant set appears as follows: For concentric intervals $I \subset \tilde{I}$, $z \in \partial(I^c)$ and $z' \in \partial \tilde{I}$ we define the probability $P_{z,z'}$ as the law of the finite-time random walk trajectory started at a uniformly distributed point in $G_N \times \{z\}$ and conditioned to exit $G_N \times \tilde{I}$ through $G_N \times \{z'\}$ at its final step. We have mentioned that the distribution of the G_N-component of X approaches the uniform distribution between different excursions from $G_N \times I$ to $(G_N \times \tilde{I})^c$. It follows that the law of these successive excursions of X under P, conditioned on the points z and z' of entrance and departure of the $\mathbb{Z}$-component, can be approximated by a product of the laws $P_{z,z'}$. This is shown in Lemma 4.3. A crucial element in the proof of the continuous-time Theorem 5.1 is the investigation of the $P_{z,z'}$-probability that a set V in the neighborhood of a point $x_{m,N}$ in $G_N \times I$ is not left vacant by one excursion. We find that up to a factor tending to 1 as N tends to infinity, this probability is equal to $\mathrm{cap}^m(\Phi_{m,N}(V))h_N/|G_N|$. With the relation between the number of such excursions taking place up to time $\alpha|G_N|^2$ and the scaled local time $\mathsf{L}^{z_{m,N}}_{\alpha|G_N|^2}/|G_N|$ from Proposition 3.3

and Lemma 3.4, the law $\mathbb{Q}^{\mathbb{G}_m \times \mathbb{Z}}_{\cdot}$, see (1.5), appears as the limiting distribution of the vacant configuration in the neighborhood of $x_{m,N}$.

Let us describe the derivation of the asymptotic behavior of the $P_{z,z'}$-probability just mentioned in a little more detail. As in [36], a key step in the proof is to show that the probability that the random walk escapes from a vertex in a set $V \subset G_N \times I$ in the vicinity of $x_{m,N}$ to the complement of $G_N \times \tilde{I}$ before hitting the set V converges to the corresponding escape probability to infinity for the set $\Phi_{m,N}(V)$ in the limit model $\mathbb{G}_m \times \mathbb{Z}$. This is where the required capacity appears. The assumption A5 that (potentially small) neighborhoods $B(y_{m,N}, r_N)$ of the points $y_{m,N}$ are isomorphic to neighborhoods in $\mathbb{G}_m$ is necessary but not sufficient for this purpose. We still need to ensure that the probability that the random walk returns from the boundary of $B(x_{m,N}, r_N)$ to the vicinity of $x_{m,N}$ before exiting $G_N \times \tilde{I}$ decays. This is the reason why we assume the existence of larger neighborhoods $C_{m,N}$ containing $B(y_{m,N}, r_N)$ in A6. These neighborhoods $C_{m,N}$ are assumed to be either identical or disjoint for points with similarly-behaved $\mathbb{Z}$-components in A8. Crucially, we assume in A7 that the sets $C_{m,N}$ are themselves isomorphic to neighborhoods in infinite graphs $\hat{\mathbb{G}}_m$ that are sufficiently close to being transient, as is formalized by A9. We additionally assume in A10 that X started from any point in the boundary of $C_{m,N} \times \mathbb{Z}$ typically does not reach the vicinity of $x_{m,N}$ until time $\lambda_N^{-1} |G_N|^\epsilon$, i.e. until well after the relaxation time of Y. These assumptions ensure that the random walk, when started from the boundary of $B(x_{m,N}, r_N)$, is unlikely to return to a point close to $x_{m,N}$ before exiting $G_N \times \tilde{I}$. For this last argument, we need the mesh size h_N of the grids to be smaller than $(\lambda_N^{-1} |G_N|^\epsilon)^{1/2}$, so that h_N can be only slightly larger than the $\lambda_N^{-1/2}$ required for the homogenization of the G_N-component.

In order to deduce Theorem 1.1 from the continuous-time result, we need an estimate on the long term-behavior of the process of jump times of X and a comparison of the local time of X and the local time of the discrete skeleton X. This requires a kind of ergodic theorem, with the feature that both time and the process itself depend on N. To show the required estimates, we use estimates on the covariance between sufficiently distant increments of the jump process that follow from bounds on the spectral gap of G_N. With the assumption (1.7), we find that the total number of jumps made by X up to a time of order $|G_N|^2$ is essentially proportional to the limit of the average weight $(1 + \beta)$ per vertex in $G_N \times \mathbb{Z}$, see Lemma 6.4. In this context, the hypothesis A1 of uniform boundedness of the vertex-weights of G_N plays an important role for stochastic domination of jump processes by homogeneous Poisson processes.

2. Notation and hypotheses

The article is organized as follows: In Section 2, we introduce notation and state the hypotheses A1-A10 for Theorem 1.1. In Section 3, we introduce the partially inhomogeneous grids with the relevant results described above. Section 4 shows that the dependence between the G_N-components of different excursions related to these grids is negligible. With these ingredients at hand, we can prove the continuous-time version of Theorem 1.1 in Section 5. The crucial estimates on the jump process needed to transfer the result to discrete time are derived in Section 6. With the help of these estimates, we finally deduce Theorem 1.1 in Section 7. Section 8 is devoted to applications of Theorem 1.1 to three concrete examples of G_N.

Throughout this article, c and c' denote positive constants changing from place to place. Numbered constants $c_0, c_1, \ldots$ are fixed and refer to their first appearance in the text. Dependence of constants on parameters appears in the notation.

Acknowledgments. The author is grateful to Alain-Sol Sznitman for proposing the problem and for helpful advice.

2. Notation and hypotheses

The purpose of this section is to introduce some useful notation and state the hypotheses A1-A10 made in Theorem 1.1.

Given any sequence a_N of real numbers, $o(a_N)$ denotes a sequence b_N with the property $b_N/a_N \to 0$ as $N \to \infty$. The notation $a \wedge b$ and $a \vee b$ is used to denote the respective minimum and maximum of the numbers a and b. For any set A, we denote by $|A|$ the number of its elements. For a set $\mathcal{B}$ of vertices in a graph $\mathcal{G}$, we denote by $\partial \mathcal{B}$ the boundary of $\mathcal{B}$, defined as the set of vertices in the complement of $\mathcal{B}$ with at least one neighbor in $\mathcal{B}$ and define the closure of $\mathcal{B}$ as $\bar{\mathcal{B}} = \mathcal{B} \cup \partial \mathcal{B}$.

We now construct the relevant probabilities for our study. For any weighted graph $\mathcal{G}$, the path space $\mathcal{P}(\mathcal{G})$ is defined as the set of right-continuous functions from $[0, \infty)$ to $\mathcal{G}$ with infinitely many discontinuities and finitely many discontinuities on compact intervals, endowed with the canonical σ-algebra generated by the coordinate projections. We let $(\mathsf{Y}_t)_{t \geq 0}$ stand for the canonical coordinate process on $\mathcal{P}(\mathcal{G})$. We consider the probability measures $P_\mathsf{y}^\mathcal{G}$ on $\mathcal{P}(\mathcal{G})$ such that Y is distributed as a continuous-time Markov chain on $\mathcal{G}$ starting from $\mathsf{y} \in \mathcal{G}$ with transition rates given by the weights $w_{\mathsf{y},\mathsf{y}'}$. Then the discrete skeleton $(Y_n)_{n \geq 0}$, defined by $Y_n = \mathsf{Y}_{\sigma_n^\mathsf{Y}}$, with $(\sigma_n^\mathsf{Y})_{n \geq 0}$ the successive times of discontinuity of Y (where $\sigma_0^\mathsf{Y} = 0$), is a random walk on $\mathcal{G}$ starting from y with transition probabilities $p^\mathcal{G}(\mathsf{y}, \mathsf{y}') = w_{\mathsf{y},\mathsf{y}'}/w_\mathsf{y}$. The discrete- and continuous-time transition probabilities for general times n and t are denoted by $p_n^\mathcal{G}(\mathsf{y}, \mathsf{y}') = P_\mathsf{y}^\mathcal{G}[Y_n = \mathsf{y}']$ and $q_t^\mathcal{G}(\mathsf{y}, \mathsf{y}') = P_\mathsf{y}^\mathcal{G}[\mathsf{Y}_t = \mathsf{y}']$. The jump process $(\eta_t^Y)_{t\geq 0}$ of Y is denoted by $\eta_t^Y = \sup\{n \geq 0 : \sigma_n^Y \leq t\}$, so that $Y_t = Y_{\eta_t^Y}, t \geq 0$.

Next, we adapt the notation of the last paragraph to the graphs we consider. Let $\mathcal{G}$ be any of the graphs $\mathbb{Z} = \{z, z', \ldots\}$ with weight $1/2$ attached to any edge, $G_N = \{y, y', \ldots\}$, $\mathbb{G}_m = \{\mathsf{y}, \mathsf{y}', \ldots\}$ or $\hat{\mathbb{G}}_m = \{\mathsf{y}, \mathsf{y}', \ldots\}$, where G_N are the finite bases of the cylinder in (1.1), and for $1 \leq m \leq M$, $\mathbb{G}_m$ are the infinite graphs in (1.3) and $\hat{\mathbb{G}}_m$ are infinite connected weighted graphs. Unlike $\mathbb{G}_m$, the graphs $\hat{\mathbb{G}}_m$ do not feature in the statement of Theorem 1.1. They do, however, play a crucial role in its proof. Indeed, we will assume that neighborhoods of the points $y_{m,N}$ that are in general much larger than $B(y_{m,N}, r_N)$ are isomorphic to subsets of $\hat{\mathbb{G}}_m$. For some examples such as the Euclidean box treated in Section 8, this assumption requires that $\hat{\mathbb{G}}_m$ be different from $\mathbb{G}_m$. Assumptions on $\hat{\mathbb{G}}_m$ will then allow us to control certain escape probabilities from the boundary of $B(x_{m,N}, r_N)$ to the complement of $G_N \times \tilde{I}$, for an interval $\tilde{I}$ containing $z_{m,N}$. See also assumptions A6-A10 and Remark 2.1 below for more on the graphs $\hat{\mathbb{G}}_m$.

Under the product measures $P_y^{\mathcal{G}} \times P_z^{\mathbb{Z}}$ on $\mathcal{P}(\mathcal{G}) \times \mathcal{P}(\mathbb{Z})$, we consider the process $X = (Y, Z)$ on $\mathcal{G} \times \mathbb{Z}$. The crucial observation is that X has the same distribution as the random walk in continuous time on $\mathcal{G} \times \mathbb{Z}$ attached to the weights

$$(2.1) \qquad w_{(y,z),(y',z')} = w_{y,y'} \mathbf{1}_{\{z=z'\}} + \frac{1}{2} \mathbf{1}_{\{y=y', |z-z'|=1\}},$$

for any pair of vertices $\{(y, z), (y', z')\}$ in $\mathcal{G} \times \mathbb{Z}$. We define the discrete skeleton $(X_n)_{n\geq 0}$ of X by $X_n = X_{\sigma_n^X}$, with $(\sigma_n^X)_{n\geq 0}$ the times of discontinuity of X (where $\sigma_0^X = 0$) and similarly $Z_n = Z_{\sigma_n^Z}$ for the times $(\sigma_n^Z)_{n\geq 0}$ of discontinuity of Z. We will often rely on the fact that

$(2.2) \qquad X$ is distributed as the random walk on $\mathcal{G} \times \mathbb{Z}$
with weights as in (2.1).

The jump process of X is defined as $\eta_t^X = \sup\{n \geq 0 : \sigma_n^X \leq t\}$. We write

$$(2.3) \qquad P_x = P_y^{G_N} \times P_z^{\mathbb{Z}}, \quad \mathbb{P}_{\mathsf{x}}^m = P_{\mathsf{y}}^{\mathbb{G}_m} \times P_z^{\mathbb{Z}} \text{ and } \hat{\mathbb{P}}_{\mathsf{x}}^m = P_{\mathsf{y}}^{\hat{\mathbb{G}}_m} \times P_z^{\mathbb{Z}},$$

for vertices $x = (y, z)$ in $G_N \times \mathbb{Z}$ and $\mathsf{x} = (\mathsf{y}, z)$ in $\mathbb{G}_m \times \mathbb{Z}$ or $\hat{\mathbb{G}}_m \times \mathbb{Z}$. Two measures on G_N are of particular interest: the reversible probability $\pi_{G_N}(y) = w_y/w(G_N)$ for $p^{G_N}(.,.)$ and the uniform measure $\mu(y) = 1/|G_N|$, $y \in G_N$, which is reversible for the continuous-time transition

2. Notation and hypotheses

probabilities $q_t^{G_N}(.,.)$, $t \geq 0$. We define

$$\tag{2.4} P^{G_N} = \sum_{y \in G_N} \mu(y) P_y^{G_N},$$

$$P_z = \sum_{y \in G_N} \mu(y) P_{(y,z)}, \text{ and}$$

$$P = \sum_{y \in G_N} \mu(y) P_{(y,0)}.$$

On any path space $\mathcal{P}(\mathcal{G})$, the canonical shift operators are denoted by $(\theta_t)_{t \geq 0}$. The shift operators for the discrete-time process X are denoted by $\theta_n^X = \theta_{\sigma_n^X}$, $n \geq 0$.

For the process X, the entrance-, exit- and hitting times of a set A are defined as

$$\tag{2.5} H_A = \inf\{n \geq 0 : X_n \in A\}, \; T_A = \inf\{n \geq 0 : X_n \notin A\}$$

$$\text{and } \tilde{H}_A = \inf\{n \geq 1 : X_n \in A\}.$$

In the case $A = \{x\}$, we simply write H_x and $\tilde{H}_x$. We also use the same notation for the corresponding times of the processes Y and Z. The analogous times for the continuous-time processes X, Y and Z are denoted H_A and T_A. Recall the definition of the local time of the $\mathbb{Z}$-projection of the random walk on $G \times \mathbb{Z}$ from (1.6). The local times of Z and its discrete skeleton Z are defined as

$$\tag{2.6} \mathsf{L}_t^z = \int_0^t \mathbf{1}_{\{\mathsf{Z}_s = z\}} ds \text{ and } \hat{L}_n^z = \sum_{l=0}^{n-1} \mathbf{1}_{\{Z_l = z\}}.$$

Note that $\hat{L}_n^z$ should not be confused with the local time L_n^z of the $\mathbb{Z}$-projection of X, defined in (1.6). The capacity of a finite subset $\mathbb{V}$ of $\mathbb{G}_m \times \mathbb{Z}$ is defined as

$$\tag{2.7} \text{cap}^m(\mathbb{V}) = \sum_{\mathbf{x} \in \mathbb{V}} \mathbb{P}_\mathbf{x}^m[\tilde{\mathsf{H}}_\mathbb{V} = \infty] w_\mathbf{x}.$$

For an arbitrary real-valued function f on G_N, the Dirichlet form $\mathcal{D}_N(f, f)$ is given by

$$\tag{2.8} \mathcal{D}_N(f, f) = \frac{1}{2} \sum_{y,y' \in G_N} (f(y) - f(y'))^2 \frac{w_{y,y'}}{|G_N|},$$

and related to the spectral gap λ_N of the continuous-time random walk Y on G_N via

$$\tag{2.9} \lambda_N = \min\left\{\frac{\mathcal{D}_N(f,f)}{\text{var}_\mu(f)} : f \text{ is not constant}\right\},$$

$$\text{where } \text{var}_\mu(f) = \mu\big((f - \mu(f))^2\big).$$

112   4. Random walks on cylinders and random interlacements

The inverse λ_N^{-1} of the spectral gap is known as the relaxation time of the continuous-time random walk, due to the estimate (4.1).

We now come to the specification of the hypotheses for Theorem 1.1. Recall that $(G_N)_{N\geq 1}$ is a sequence of finite connected weighted graphs. We consider $M \geq 1$, sequences $x_{m,N} = (y_{m,N}, z_{m,N})$, $1 \leq m \leq M$, in $G_N \times \mathbb{Z}$ and an $0 < \epsilon < 1$ such that the assumptions A1-A10 below hold. The first assumption is that the weights attached to vertices of G_N are uniformly bounded from above and below, i.e.

(A1)   there are constants $0 < c_0 \leq c_1$ such that $c_0 \leq w_y \leq c_1$,
        for all $y \in G_N$.

A frequently used consequence of this assumption is that the jump process of Y under P^G can be bounded from above and from below by a Poisson process of constant parameter, see Lemma 2.4 below. Moreover, by taking a function f vanishing everywhere except at a single vertex in (2.9), A1 implies that $\lambda_N \leq c$. If in addition also the edge-weights $w_{y,y'}$ of G_N are uniformly elliptic, it follows from Cheeger's inequality (see [**28**], Lemma 3.3.7, p. 383) that the relaxation time λ_N^{-1} is bounded from above by $c|G_N|^2$. We assume a little bit more, namely that for ϵ as above,

(A2) $$\lambda_N^{-1} \leq |G_N|^{2-\epsilon},$$

which in particular rules out nearly one-dimensional graphs G_N. We further assume that the mutual distances between different sequences $x_{m,N}$ diverge,

(A3) $$\lim_N \min_{1 \leq m < m' \leq M} d(x_{m,N}, x_{m',N}) = \infty,$$

and that in scale $|G_N|$, the $\mathbb{Z}$-components of the sequences $z_{m,N}$ converge:

(A4) $$\lim_N \frac{z_{m,N}}{|G_N|} = v_m \in \mathbb{R}, \text{ for } 1 \leq m \leq M.$$

The key assumption is the existence of balls of diverging size centered at the points $y_{m,N}$ that are isomorphic to balls with fixed centers o_m in the infinite graphs $\mathbb{G}_m$:

(A5)   For some $r_N \to \infty$, there are isomorphisms $\phi_{m,N}$ from
        $B(y_{m,N}, r_N)$ to $B(o_m, r_N) \subset \mathbb{G}_m$, such that $\phi_{m,N}(y_{m,N}) = o_m$
        for all N, m.

In the proof of Theorem 1.1, we want to show the decay of the probability that the random walk X under P returns to the close vicinity of the center $x_{m,N}$ from the boundary of each of the balls $B(x_{m,N}, r_N) \subset G_N \times \mathbb{Z}$ before exiting a large box. With this aim in mind, we make the

2. Notation and hypotheses

remaining assumptions. For any m, N, we assume that there exists an associated subset $C_{m,N}$ of G_N such that

(A6) $$B(y_{m,N}, r_N) \subseteq C_{m,N},$$

and $\bar{C}_{m,N}$ are isomorphic to a subset of the auxiliary limit model $\hat{\mathbb{G}}_m$, i.e.

(A7) there is an isomorphism $\psi_{m,N}$ from $\bar{C}_{m,N}$ with a set $\bar{\mathbb{C}}_m \subset \hat{\mathbb{G}}_m$, such that $\psi_{m,N}(\partial C_{m,N}) = \partial \mathbb{C}_{m,N}$,

where the last condition is to ensure that the distributional identity (2.13) below holds. Note that we are allowing the infinite graphs $\hat{\mathbb{G}}_m$ to be different from $\mathbb{G}_m$. For an explanation, we refer to Remark 2.1 below (see also Remark 8.4). We further assume that the sets $C_{m,N}$ as m varies are essentially either disjoint or equal (unless the corresponding $\mathbb{Z}$-components $z_{m,N}$ are far apart), i.e.

(A8) whenever $v_m = v_{m'}$, then for all N either $C_{m,N} = C_{m',N}$, or $C_{m,N} \cap C_{m',N} = \emptyset$.

Concerning the limit model $\hat{\mathbb{G}}_m$, we require that the measure of a constant-size ball centered at $\hat{o}_{m,N} \stackrel{(\text{def.})}{=} \psi_{m,N}(y_{m,N})$ under the law $Y_n \circ P^{\hat{\mathbb{G}}_m}$ decays faster than $n^{-\frac{1}{2}-\epsilon}$,

(A9) $$\lim_{n \to \infty} n^{\frac{1}{2}+\epsilon} \sup_{\mathbf{y}_0 \in \hat{\mathbb{G}}_m} \sup_{\mathbf{y} \in B(\hat{o}_{m,N}, \rho_0), N \geq 1} p_n^{\hat{\mathbb{G}}_m}(\mathbf{y}_0, \mathbf{y}) = 0, \text{ for any } \rho_0 > 0.$$

This assumption is only used to prove Lemma 2.3 below. Let us mention that A9 typically holds whenever the on-diagonal transition densities decay at the same rate, see Remark 2.2 below. Finally, we assume that the random walk on $G_N \times \mathbb{Z}$, started at the interior boundary of $C_{m,N} \times \mathbb{Z}$, is unlikely to reach the vicinity of $x_{m,N}$ until well after the relaxation time of Y:

(A10) $$\lim_N \sup_{\mathbf{y}_0 \in \partial(C_m^c), z_0 \in \mathbb{Z}} P_{(y_0, z_0)}[H_{(\phi_{m,N}^{-1}(\mathbf{y}), z_{m,N}+z)} < \lambda_N^{-1} |G_N|^\epsilon] = 0,$$

for any $(\mathbf{y}, z) \in \mathbb{G}_m \times \mathbb{Z}$ (note that $\phi_{m,N}^{-1}(\mathbf{y})$ is well-defined for large N by A5).

REMARK 2.1. The infinite graphs $\hat{\mathbb{G}}_m$ in A7 can be different from the graphs $\mathbb{G}_m$ describing the neighborhoods of the points $y_{m,N}$. The reason is that for A10 to hold, the sets $C_{m,N}$ will generally have to be of much larger diameter than their subsets $B(y_m, r_N)$. Hence, $\bar{C}_m$ is not necessarily isomorphic to a subset of the same infinite graph as $B(y_m, r_N)$. This situation occurs, for example, if G_N is given by a Euclidean box, see Remark 8.4.

REMARK 2.2. Typically, the weights attached to the vertices of $\hat{\mathbb{G}}_m$ are uniformly bounded from above and from below, as are the weights in G_N (see (A1)). In this case, assumption A9 holds in particular whenever one has the on-diagonal decay

$$\lim_n n^{\frac{1}{2}+\epsilon} \sup_{y \in \hat{\mathbb{G}}_m} p_n^{\hat{\mathbb{G}}_m}(y,y) \to 0,$$

see [**42**], Lemma 8.8, p. 108, 109.

From now on, we often drop the N from the notation in G_N, $C_{m,N}$, $x_{m,N}$, $\phi_{m,N}$ and $\psi_{m,N}$. We extend the isomorphisms ϕ_m and ψ_m in A5 and A7 to isomorphisms Φ_m and $\Psi_m^{z_0}$ defined on $B(y_m, r_N) \times \mathbb{Z}$ and on $\bar{C}_m \times \mathbb{Z}$ by

(2.10) $\qquad \Phi_m : (y,z) \mapsto (\phi_m(y), z - z_m),$ and

(2.11) $\qquad \Psi_m^{z_0} : (y,z) \mapsto (\psi_m(y), z - z_0),$ for $z_0 \in \mathbb{Z}.$

A crucial consequence of (A5) and (A7) is that for $r_N \geq 1$,

(2.12) $\quad (\mathsf{X}_t : 0 \leq t \leq \mathsf{T}_{B(y_m, r_N-1) \times \mathbb{Z}})$ under P_x has the same distribution as $(\Phi_m^{-1}(\mathsf{X}_t) : 0 \leq t \leq \mathsf{T}_{B(o_m, r_N-1) \times \mathbb{Z}})$ under $\mathbb{P}^m_{\Phi_m(x)}$, and

(2.13) $\quad (\mathsf{X}_t : 0 \leq t \leq \mathsf{T}_{\mathbb{C}_m \times \mathbb{Z}})$ under P_x has the same distribution as $((\Psi_m^{z_0})^{-1}(\mathsf{X}_t) : 0 \leq t \leq \mathsf{T}_{\mathbb{C}_m \times \mathbb{Z}})$ under $\hat{\mathbb{P}}^m_{\Psi_m^{z_0}(x)}$.

The assumption A9 only enters the proof of the following lemma showing the decay of the probability that the random walk on the cylinders $\mathbb{G}_m \times \mathbb{Z}$ or $\hat{\mathbb{G}}_m \times \mathbb{Z}$ returns from distance ρ to a constant-size neighborhood of $(o_m, 0)$ or $(\psi_m(y_m), 0)$ as ρ tends to infinity. Note that this in particular implies that these cylinders are transient and the random interlacements appearing in Theorem 1.1 make sense.

LEMMA 2.3. $(1 \leq m \leq M)$ Assuming A1-A10, for any $\rho_0 > 0$,

(2.14) $\qquad \lim_{\rho \to \infty} \sup_{\substack{d(\mathsf{x}, (\hat{o}_m, 0)) \leq \rho_0 \\ d(\mathsf{x}_0, \mathsf{x}) \geq \rho}} \hat{\mathbb{P}}^m_{\mathsf{x}_0}[H_\mathsf{x} < \infty] = 0,$ and

$\qquad \lim_{\rho \to \infty} \sup_{\substack{d(\mathsf{x}, (o_m, 0)) \leq \rho_0 \\ d(\mathsf{x}_0, \mathsf{x}) \geq \rho}} \mathbb{P}^m_{\mathsf{x}_0}[H_\mathsf{x} < \infty] = 0.$

The proof of Lemma 2.3 requires the following two lemmas of frequent use.

LEMMA 2.4. Let $\mathcal{G}$ be a weighted graph such that

$$0 < \inf_y w_y \leq \sup_y w_y < \infty.$$

2. Notation and hypotheses

(2.15)   Under $P_y^{\mathcal{G}}$, $e_n = (\sigma_n^Y - \sigma_{n-1}^Y)w_{Y_{n-1}}$, $n \geq 1$, is a sequence of iid $\exp(1)$ random variables, independent of Y, and

(2.16) $\eta_t^{\inf_y w_y} \leq \eta_t^Y \leq \eta_t^{\sup_y w_y}$, for $t \geq 0$,

where $\eta_t^\nu = \sup\{n \geq 0 : e_1 + \ldots + e_n \leq \nu t\}$, $t \geq 0$, with $(e_n)_{n\geq 1}$ as defined above, is a Poisson process with rate $\nu \geq 0$.

PROOF. The assertion (2.15) follows from a standard construction of the continuous-time Markov chain Y, see for example [23], pp. 88, 89. For (2.16), note that for any $k \geq 0$,

(2.17) $$\frac{w_{Y_k}}{\sup_y w_y} \leq 1 \leq \frac{w_{Y_k}}{\inf_y w_y},$$

hence for $t \geq 0$,

$$\eta_t^Y = \sup\left\{n \geq 0 : \sum_{k=1}^n (\sigma_k^Y - \sigma_{k-1}^Y) \leq t\right\}$$

$$\stackrel{(2.17)}{\leq} \sup\left\{n \geq 0 : \sum_{k=1}^n (\sigma_k^Y - \sigma_{k-1}^Y)\frac{w_{Y_{k-1}}}{\sup_y w_y} \leq t\right\} = \eta_t^{\sup_y w_y},$$

as well as

$$\eta_t^Y \stackrel{(2.17)}{\geq} \sup\left\{n \geq 0 : \sum_{k=1}^n (\sigma_k^Y - \sigma_{k-1}^Y)\frac{w_{Y_{k-1}}}{\inf_y w_y} \leq t\right\} = \eta_t^{\inf_y w_y}.$$

$\square$

LEMMA 2.5.

(2.18) $P_z^{\mathbb{Z}}[z' \in \mathsf{Z}_{[s,t]}] \leq c\dfrac{1+t-s}{\sqrt{s}}$, for $0 < s \leq t < \infty$, $z, z' \in \mathbb{Z}$.

PROOF. By the strong Markov property applied at time $s + \mathsf{H}_{z'} \circ \theta_s$,

(2.19) $$E_z^{\mathbb{Z}}\left[\int_s^{t+1} \mathbf{1}_{\{Z_r = z'\}} dr\right]$$

$$\geq E_z^{\mathbb{Z}}\left[s + \mathsf{H}_{z'} \circ \theta_s \leq t, \int_{s+\mathsf{H}_{z'}\circ\theta_s}^{t+1} \mathbf{1}_{\{Z_r = z'\}} dr\right]$$

$$\geq P_z^{\mathbb{Z}}[\mathsf{H}_{z'} \circ \theta_s \leq t - s] E_{z'}^{\mathbb{Z}}\left[\int_0^1 \mathbf{1}_{\{Z_r = z'\}} dr\right]$$

$$\geq P_z^{\mathbb{Z}}[z' \in \mathsf{Z}_{[s,t]}] E_{z'}^{\mathbb{Z}}[\sigma_1^Z \wedge 1] \geq c P_z^{\mathbb{Z}}[z' \in \mathsf{Z}_{[s,t]}].$$

It follows from the local central limit theorem, see [19], (1.10), p. 14, (or from a general upper bound on heat kernels of random walks, see Corollary 14.6 in [47]) that

(2.20) $P_z^{\mathbb{Z}}[Z_n = z'] \leq c/\sqrt{n}$, for all z and z' in $\mathbb{Z}$ and $n \geq 1$.

Using an exponential bound on the probability that a Poisson variable of intensity $2t$ is not in the interval $[t, 4t]$, it readily follows that $P_z^{\mathbb{Z}}[Z_t = z'] \leq c/\sqrt{t}$ for all $t > 0$, hence

$$E_z^{\mathbb{Z}}\Big[\int_s^{t+1} 1_{\{Z_r = z'\}} dr\Big] \leq c \int_s^{t+1} \frac{1}{\sqrt{r}} dr \leq c \frac{1+t-s}{\sqrt{s}}.$$

With (2.19), this implies (2.18). □

PROOF OF LEMMA 2.3. Denote by $\mathbb{G}$ either one of the graphs $\hat{\mathbb{G}}_m$ or $\mathbb{G}_m$ and by $\mathbb{P}$ the corresponding probabilities $\hat{\mathbb{P}}^m$ and $\mathbb{P}^m$. Assume for the moment that for all $n \geq c(\epsilon, \rho_0)$,

(2.21) $$\sup_{y_0 \in \mathbb{G}} \sup_{y \in B(o, \rho_0)} p_n^{\mathbb{G}}(y_0, y) \leq c(\rho_0) n^{-\frac{1}{2} - \epsilon},$$

where o denotes the corresponding vertex $\hat{o}_{m,N}$ or o_m. For any points $\mathbf{x} = (\mathbf{y}, z)$ in $B((o, 0), \rho_0)$ and $\mathbf{x}_0 = (\mathbf{y}_0, z_0)$ in $\mathbb{G} \times \mathbb{Z}$ such that $d(\mathbf{x}_0, \mathbf{x}) \geq \rho$, we have

(2.22) $$\mathbb{P}_{\mathbf{x}_0}[H_{\mathbf{x}} < \infty] \leq \sum_{n=[\rho]}^{\infty} P_{(\mathbf{y}_0, z_0)}[Y_n = \mathbf{y}, z \in \mathbb{Z}_{[\sigma_n^Y, \sigma_{n+1}^Y]}],$$

By independence of (Y, σ^Y) and Z, the probability in this sum can be rewritten as

$$E_{\mathbf{y}_0}^{\mathbb{G}}\Big[Y_n = \mathbf{y}, P_{z_0}^{\mathbb{Z}}[z \in \mathbb{Z}_{[s,t]}]\Big|_{\substack{s = \sigma_n^Y \\ t = \sigma_{n+1}^Y}}\Big],$$

which by the estimate (2.18) and the strong Markov property at time σ_n^Y is smaller than

$$cE_{\mathbf{y}_0}^{\mathbb{G}}\Big[Y_n = \mathbf{y}, \frac{1 + \sigma_1 \circ \theta_{\sigma_n^Y}}{\sqrt{\sigma_n^Y}}\Big] \overset{(A1)}{\leq} cE_{\mathbf{y}_0}^{\mathbb{G}}\Big[Y_n = \mathbf{y}, \frac{1}{\sqrt{\sigma_n^Y}}\Big].$$

By (2.15) and A1, the sum in (2.22) can be bounded by

(2.23) $$c \sum_{n=[\rho]}^{\infty} p_n^{\mathbb{G}}(\mathbf{y}_0, \mathbf{y}) E\Big[\frac{1}{\sqrt{e_1 + \ldots + e_n}}\Big] \leq c \sum_{n=[\rho]}^{\infty} p_n^{\mathbb{G}}(\mathbf{y}_0, \mathbf{y}) \frac{1}{\sqrt{n}},$$

where we have used that $E[1/(e_1 + \ldots + e_n)] = 1/(n-1)$ for $n \geq 2$ (note that $e_1 + \ldots + e_n$ is $\Gamma(n, 1)$-distributed), together with Jensen's inequality. By the bound assumed in (2.21), this implies with (2.22) that

$$\sup_{\substack{d(\mathbf{x},(o,0)) \leq \rho_0 \\ d(\mathbf{x}_0, \mathbf{x}) \geq \rho}} \mathbb{P}_{\mathbf{x}_0}[H_{\mathbf{x}} < \infty] \leq c(\rho_0) \sum_{n=[\rho]}^{\infty} n^{-1-\epsilon}.$$

Since the right-hand side tends to 0 as ρ tends to infinity, this proves both claims in (2.14), provided (2.21) holds for $\hat{\mathbb{G}}_m$ and $\mathbb{G}_m$ in place of $\mathbb{G}$. In fact, (2.21) does hold for $\mathbb{G} = \hat{\mathbb{G}}_m$ by assumption A9, and also holds for $\mathbb{G} = \mathbb{G}_m$ by the following argument: Consider any $\mathbf{y}_0 \in$

3. Auxiliary results on excursions and local times   117

$\mathbb{G}_m$, $\mathbf{y} \in B(o_m, \rho_0)$ and $n \geq 0$. Choose N sufficiently large such that $r_N - d(\mathbf{y}_0, o_m) > n$ and both $\mathbf{y}_0$ and $\mathbf{y}$ are contained in $B(o_m, r_N)$ (cf. A5). Using the isomorphism $\hat{\psi} = \psi_m \circ \phi_m^{-1}$ from $B(o_m, r_N)$ to $B(\hat{o}_m, r_N) \subset \hat{\mathbb{G}}_m$, we deduce that

$$(2.24) \quad p_n^{\mathbb{G}_m}(\mathbf{y}_0, \mathbf{y}) = P_{\mathbf{y}_0}^{\mathbb{G}_m}[Y_n = \mathbf{y}, T_{B(o_m, r_N - 1)} \geq r_N - d(\mathbf{y}_0, o_m)].$$
$$= P_{\hat{\psi}(\mathbf{y}_0)}^{\hat{\mathbb{G}}_m}[Y_n = \hat{\psi}(\mathbf{y}), T_{B(\hat{o}_m, r_N - 1)} \geq r_N - d(\mathbf{y}_0, o_m)]$$
$$\leq p_n^{\hat{\mathbb{G}}_m}(\hat{\psi}(\mathbf{y}_0), \hat{\psi}(\mathbf{y})) \leq c(\rho_0) n^{-\frac{1}{2} - \epsilon},$$

using A9 in the last step. This concludes the proof of Lemma 2.3. □

3. Auxiliary results on excursions and local times

In this section we reproduce a suitable version of the partially inhomogeneous grids on $\mathbb{Z}$ introduced in Section 2 of [36]. These grids allow to relate excursions of the walk Z associated to the grid points to the total time elapsed and to the local time $\hat{L}$ of Z. This is essentially the content of Proposition 3.3 below, quoted from [36]. We then complement this result with an estimate relating the local time $\hat{L}$ of Z to the local time L of the continuous-time process Z in Lemma 3.4.

For integers $1 \leq d_N \leq h_N$ and points $z_{l,N}^*$, $1 \leq l \leq M$, in $\mathbb{Z}$ (to be specified below), we define the intervals

$$(3.1) \qquad I_l = [z_l^* - d_N, z_l^* + d_N] \subseteq \tilde{I}_l = (z_l^* - h_N, z_l^* + h_N),$$

dropping the N from $z_{l,N}^*$ for ease of notation. The collections of these intervals are denoted by

$$(3.2) \qquad \mathcal{I} = \{I_l, 1 \leq l \leq M\}, \text{ and } \tilde{\mathcal{I}} = \{\tilde{I}_l, 1 \leq l \leq M\}.$$

The anisotropic grid $\mathcal{G}_N \subset \mathbb{Z}$, is defined as in [36], (2.4):

$$(3.3) \qquad \mathcal{G}_N = \mathcal{G}_N^* \cup \mathcal{G}^0, \text{ where } \mathcal{G}_N^* = \{z_l^*, 1 \leq l \leq M\} \text{ and}$$
$$\mathcal{G}_N^0 = \{z \in 2h_N \mathbb{Z} : |z - z_l^*| \geq 2h_N, \text{ for } 1 \leq l \leq M\}.$$

It remains to choose d_N, h_N and z_l^*. In [36], no upper bound other than $o(|\mathcal{G}_N|)$ is needed on the distance between neighboring grid points, but we want an upper bound not much larger than $\lambda_N^{-1/2}$. A consequence of this requirement is that unlike in [36], we may attach several points z_l^* to the same limit v_m in A4. We satisfy this requirement by a judicious choice such that

$$(3.4) \qquad \lambda_N^{-1/2}|\mathcal{G}_N|^{\epsilon/8} \leq d_N, \quad d_N = o(h_N), \quad h_N \leq \lambda_N^{-1/2}|\mathcal{G}_N|^{\epsilon/4},$$

$$(3.5) \qquad \min_{1 \leq l < l' \leq M} |z_l^* - z_{l'}^*| \geq 100 h_N, \text{ and}$$

$$(3.6) \qquad \{z_1, \ldots, z_M\} \subseteq \cup_{l=1}^M [z_l^* - [d_N/2], z_l^* + [d_N/2]],$$

for all $N \geq c(\epsilon, M)$.

118   4. Random walks on cylinders and random interlacements

PROPOSITION 3.1. *Points $z_1^*, \ldots, z_M^*$ in $\mathbb{Z}$ and sequences d_N, h_N in $\mathbb{N}$ satisfying (3.4)-(3.6) exist.*

The proof of Proposition 3.1 is a consequence of the following simple lemma, asserting that for prescribed numbers a, b and $q \geq 2$, any M points in a metric space can be covered by balls of radius between a and $b^{2M}a$, such that the balls with radius multiplied by b are disjoint and no more than M balls are required.

LEMMA 3.2. *Let $\mathcal{X}$ be a metric space and $x_1, \ldots, x_M$, $M \geq 1$, points in $\mathcal{X}$. Consider real numbers $a \geq 1$ and $b \geq 2$. Then for some $M_* \leq M$ and $a \leq p \leq b^{2M}a$, there are points $\{x_1^*, \ldots, x_{M_*}^*\}$ in $\mathcal{X}$ such that*

$$\bigcup_{1 \leq i \leq M_*} B(x_i^*, p) \supseteq \{x_1, \ldots, x_M\},$$

and the balls $(B(x_i^*, bp))_{i=1}^{M_*}$ are disjoint,

where $B(x, r)$ denotes the closed ball of radius $r \geq 0$ centered at $x \in \mathcal{X}$.

PROOF OF PROPOSITION 3.1. Lemma 3.2, applied with $\mathcal{X} = \mathbb{Z}$ and the points $z_1, \ldots, z_M$ with

$$a = [\lambda_N^{-1/2}|G|^{\epsilon/8}] \text{ and } b = [(|G|^{\epsilon/8})^{1/(2M+1)}],$$

yields points $z_1^*, \ldots, z_{M_*}^*$ in $\mathbb{Z}$ and a p between a and $b^{2M}a$ such that (3.4)-(3.6) hold for $d_N = [2p]$, $h_N = [bp/100]$ and M_* in place of M. The additional points $z_{M_*+1}^*, \ldots, z_M^*$ can be chosen arbitrarily subject only to (3.5). □

PROOF OF LEMMA 3.2. For $m \geq 0$, set

$$k_m = \min\{k \geq 0 : \text{ for some } x_1', \ldots, x_k' \text{ in } \mathcal{X},$$
$$\cup_{i=1}^k B(x_i', b^{2m}a) \supseteq \{x_1, \ldots, x_M\}\},$$

and denote points for which the minimum is attained by $x_1^m, \ldots, x_{k_m}^m$. The first observation on k_m is that clearly $1 \leq k_m \leq M$. The second observation is that for $m \geq 0$,

either the balls $B(x_i^m, b^{2m+1}a)$, $1 \leq i \leq k_m$, are disjoint, or $k_{m+1} < k_m$.

Indeed, assume that $\bar{x} \in B(x_i^m, b^{2m+1}a) \cap B(x_j^m, b^{2m+1}a)$ for $1 \leq i < j \leq k_m$. Then since $b \geq 2$, the $k_m - 1$ balls of radius $b^{2(m+1)}a$ centered at $(\{x_1^m, \ldots, x_{k_m}^m\} \cup \{\bar{x}\}) \setminus \{x_i^m, x_j^m\}$ still cover $\{x_1, \ldots, x_M\}$. Thanks to these two observations, we may define

$$m_* = \min\{m \geq 0 : B(x_i^m, b^{2m+1}a), 1 \leq i \leq k_m, \text{ are disjoint}\} \leq M,$$

and set $M_* = k_{m_*}$, $x_i^* = x_i^{m_*}$ for $1 \leq i \leq M_*$ and $p = b^{2m_*}a$. □

The grids $\mathcal{G}_N$ we consider from now on are specified by (3.1)-(3.6). In order to define the associated excursions, we define the sets C and

3. Auxiliary results on excursions and local times

O, whose components are intervals of radius d_N and h_N, centered at the points in the grid $\mathcal{G}_N$, i.e.

(3.7) $\quad C = \mathcal{G}_N + [-d_N, d_N] \subset O = \mathcal{G}_N + (-h_N, h_N).$

The times R_n and D_n of return to C and departure from O of the process Z are defined as

(3.8) $\quad R_1 = H_C, D_1 = T_O \circ \theta_{R_1} + R_1,$ and for $n \geq 1,$
$$R_{n+1} = R_1 \circ \theta_{D_n} + D_n, \quad D_{n+1} = D_1 \circ \theta_{D_n} + D_n,$$

so that $0 \leq R_1 < D_1 < \ldots < R_n < D_n$, $P_z^\mathbb{Z}$-a.s. For later use, we denote for any $\alpha > 0$,

(3.9) $\quad \mathsf{t}_N = E_0^\mathbb{Z}[T_{(-h_N+d_N, h_N-d_N)}] + E_{d_N}^\mathbb{Z}[T_{(-h_N, h_N)}]$
$$= (h_N - d_N)^2 + h_N^2 - d_N^2,$$

(3.10) $\quad \sigma_N = [\alpha |G|^2 / \mathsf{t}_N], \quad k_*(N) = \sigma_N - [\sigma_N^{3/4}], \quad k^*(N) = \sigma_N + [\sigma_N^{3/4}],$

where we will often drop the N from now on. We come to the crucial result on these returns and departures from [36], relating the times D_k to the total time elapsed (3.11) and to the local time $\hat{L}$ of Z ((3.12)-(3.14)).

PROPOSITION 3.3. *Assuming A2,*

(3.11) $\quad \lim_N P_0^\mathbb{Z}[D_{k_*} \leq \alpha |G_N|^2 \leq D_{k^*}] = 1.$

(3.12) $\quad \limsup_N \sup_{z \in C} E_0^\mathbb{Z}[(|\hat{L}_{[\alpha |G_N|^2]}^z - \hat{L}_{D_{k_*}}^z|/|G_N|) \wedge 1] = 0.$

(3.13) $\quad \sup_N \max_{I \in \mathcal{I}} \frac{h_N}{|G_N|} E_0^\mathbb{Z} \Big[\sum_{1 \leq k \leq k_*} 1_{\{Z_{R_k} \in I\}} \Big] < \infty.$

(3.14) $\quad \lim_N \max_{I \in \mathcal{I}} \sup_{z \in I} E_0^\mathbb{Z} \Big[\Big| \hat{L}_{D_{k_*}}^z - h_N \sum_{1 \leq k \leq k_*} 1_{\{Z_{R_k} \in I\}} \Big| \Big] / |G_N| = 0.$

PROOF. The above statement is proved by Sznitman in [36]. Indeed, in [36], the author considers three sequences of non-negative integers $(a_N)_{N \geq 1}, (h_N)_{N \geq 1}, (d_N)_{N \geq 1}$, such that

(3.15) $\quad \lim_N a_N = \lim_N h_N = \infty,$ and
$\quad d_N = o(h_N), h_N = o(a_N)$ (cf. (2.1) in [36]),

as well as sequences $z_{l,N}^*$ of points in $\mathbb{Z}$ satisfying (3.5) (cf. (2.2) in [36]). The grids $\mathcal{G}_N$ are then defined as in (3.3) (cf. (2.4) in [36]) and the corresponding sets C and O as in (3.7) (cf. (2.5) in [36]). For any $\gamma \in (0,1], z \in \mathbb{Z}$, Sznitman in [36] then introduces the canonical law Q_z^γ on $\mathbb{Z}^\mathbb{N}$ of the random walk on $\mathbb{Z}$ which jumps to one of its two neighbors with probability $\gamma/2$ and stays at its present location with probability $1 - \gamma$. The times $(R_n)_{n \geq 1}$ and $(D_n)_{n \geq 0}$ of return to C and departure from O are introduced in (2.9) of [36], exactly as in (3.8) above. The sequences $\mathsf{t}_N, \sigma_N, k_*(N), k^*(N)$ are defined in (2.10)-(2.12) of [36] as in (3.9) and (3.10) above, with $|G_N|$ replaced by a_N and $E^{\mathbb{Z}}$ replaced by the Q^γ-expectation E^γ. Under these conditions, the statements (3.11)-(3.14) are proved in [**36**], Proposition 2.1, with $|G_N|$ replaced by a_N and $P_0^{\mathbb{Z}}$ and $E_0^{\mathbb{Z}}$ replaced by P_0^γ and E_0^γ. All we have to do to deduce the above statements is to choose $\gamma = 1$ and $a_N = |G_N|$ in Proposition 2.1 of [**36**], noting that (3.15) is then satisfied, by (3.4) and A2. □

We now relate the local time of Z to the local time of the continuous-time process Z.

LEMMA 3.4.

(3.16) $\quad \sup_{z \in \mathbb{Z}} E_0^{\mathbb{Z}}[\hat{L}_{[\alpha|G_N|^2]}^z] \leq c(\alpha)|G_N|$, for $\alpha > 0$.

(3.17) $\quad \limsup_N \sup_{z \in \mathbb{Z}} E_0^{\mathbb{Z}}[(|\mathsf{L}_{\alpha|G_N|^2}^z - \hat{L}_{[\alpha|G_N|^2]}^z|/|G_N|) \wedge 1] = 0$.

PROOF. For (3.16), apply the bound $P_0[Z_n = z] \leq c/\sqrt{n}$ (cf. (2.20)), see (2.34) in [**36**].

We write $T = \alpha |G|^2$. By the strong Markov property applied at time $\sigma_{[T]}^\mathsf{Z} \wedge T$,

(3.18) $\quad E_0^{\mathbb{Z}}[|\mathsf{L}_{\sigma_{[T]}^\mathsf{Z}}^z - \mathsf{L}_T^z|] = E_0^{\mathbb{Z}}\left[\int_{\sigma_{[T]}^\mathsf{Z} \wedge T}^{\sigma_{[T]}^\mathsf{Z} \vee T} \mathbf{1}_{\{\mathsf{Z}_s = z\}} ds\right]$

$\leq \sup_{z_0 \in \mathbb{Z}} E_{z_0}^{\mathbb{Z}}\left[\int_0^{|\sigma_{[T]}^\mathsf{Z} - T|} \mathbf{1}_{\{\mathsf{Z}_s = z\}} ds\right]$

$\leq \int_0^{T^{2/3}} \sup_{z_0 \in \mathbb{Z}} P_{z_0}^{\mathbb{Z}}[\mathsf{Z}_s = z] ds + E_0^{\mathbb{Z}}[(\sigma_{[T]}^\mathsf{Z} - T)^2]/T^{2/3}$,

using the Chebyshev inequality in the last step. By the bound (2.18) on $P_{z_0}^{\mathbb{Z}}[\mathsf{Z}_s = z]$ and a bound of cT on the variance of the $\Gamma([T], 1)$-distributed variable $\sigma_{[T]}^\mathsf{Z}$, the right-hand side of (3.18) is bounded by $cT^{1/3}$. Hence, the expectation in (3.17) is bounded by

(3.19) $\quad c(\alpha)|G|^{-1/3} + E_0^{\mathbb{Z}}[(|\mathsf{L}_{\sigma_{[T]}^\mathsf{Z}}^z - \hat{L}_{[T]}^z|/|G|) \wedge 1]$.

The strategy is to now split up the last expectation into expectations on the events

$A_1 = \{\delta|G| \leq \hat{L}_{[T]}^z \leq \theta|G|\}$, $A_2 = \{\hat{L}_{[T]}^z < \delta|G|\}$, $A_3 = \{\hat{L}_{[T]}^z > \theta|G|\}$,

for $0 < \delta < \theta$. In this way, one obtains the following bound on (3.19):

(3.20) $\quad c(\alpha)|G|^{-1/3}$

$+ E_0^{\mathbb{Z}}\left[A_1, \left(\left|\sum_{n=0}^{[T]-1}(\sigma_{n+1}^\mathsf{Z} - \sigma_n^\mathsf{Z} - 1)\mathbf{1}_{\{Z_n = z\}}\right|/|G|\right) \wedge 1\right]$

$+ 2\delta + P_0^{\mathbb{Z}}[A_3]$,

3. Auxiliary results on excursions and local times

where we have used the fact that $(\sigma^Z_{n+1} - \sigma^Z_n)_{n\geq 0}$ are iid $\exp(1)$ variables independent of Z to bound the expectation on A_2 by 2δ. By Chebyshev's inequality and (3.16),

$$P^Z_0[A_3] \leq E^Z_0[\hat{L}^z_{[\alpha|G|^2]}]/(\theta|G|) \leq c(\alpha)/\theta.$$

In order to bound the expectation in (3.20), we apply Fubini's theorem to obtain

$$E^Z_0\left[A_1, \left(\left|\sum_{n=0}^{[T]-1}(\sigma^Z_{n+1} - \sigma^Z_n - 1)\mathbf{1}_{\{Z_n=z\}}\right|/|G|\right) \wedge 1\right]$$

$$\leq E^Z_0\left[A_1, f(\hat{L}^z_{[T]})\frac{\hat{L}^z_{[T]}}{|G|}\right],$$

where for any $l \geq 1$, $f(l) = E^Z_0\left[\left(\left|\sum_{n=0}^{l-1}(\sigma^Z_{n+1} - \sigma^Z_n - 1)\right|/l\right) \wedge (|G|/l)\right]$.

Collecting the above estimates and using the definition of A_1, we have found the following bound on the expectation in (3.17) for any $z \in \mathbb{Z}$:

$$c(\alpha)|G|^{-1/3} + \theta \sup_{l \geq \delta|G|} f(l) + 2\delta + \frac{c(\alpha)}{\theta}.$$

Note that this expression does not depend on z, so it remains unchanged after taking the supremum over all $z \in \mathbb{Z}$. Since moreover $\sup_{l\geq \delta|G|} f(l)$ tends to 0 as $|G|$ tends to infinity by the law of large numbers and dominated convergence, this shows that the left-hand side of (3.17) (with lim replaced by lim sup) is bounded from above by $2\delta + c(\alpha)/\theta$. The result follows by letting δ tend to 0 and θ to infinity. □

Consider now the times R_n and D_n, defined as the continuous-time analogs of the times R_n and D_n in (3.8):

$$\mathsf{R}_n = \sigma^Z_{R_n} \text{ and } \mathsf{D}_n = \sigma^Z_{D_n}, \text{ for } n \geq 1,$$

so that the times R_n and D_n coincide with the successive times of return to C and departure from O for the process Z. We record the following observation:

LEMMA 3.5. *For any sequence $a_N \geq 0$ diverging to infinity,*

(3.21) $$\limsup_N \sup_{z\in\mathbb{Z}} E^Z_z[|\mathsf{D}_{a_N}/D_{a_N} - 1| \wedge 1] = 0.$$

PROOF. We define the function $g : \mathbb{N} \to \mathbb{R}$ by $g(n) = \sum_{i=1}^n (\sigma^Z_i - \sigma^Z_{i-1})/n$, so that $\mathsf{D}_{a_N}/D_{a_N} = g(D_{a_N})$. By independence of the two sequences $(\sigma_n^{\mathbb{Z}})_{n \geq 1}$ and $(D_n)_{n \geq 1}$, Fubini's theorem yields

(3.22) $$\sup_{z \in \mathbb{Z}} E_z^{\mathbb{Z}}[|\mathsf{D}_{a_N}/D_{a_N} - 1| \wedge 1]$$
$$= \sup_{z \in \mathbb{Z}} E_z^{\mathbb{Z}}\big[E_0^{\mathbb{Z}}[|g(n) - 1| \wedge 1]\big|_{n=D_{a_N}}\big],$$

where we have used that the distribution of $(\sigma_n^{\mathbb{Z}})_{n \geq 1}$ is the same under all measures $P_z^{\mathbb{Z}}$, $z \in \mathbb{Z}$. Fix any $\epsilon > 0$. By the law of large numbers, the $E_0^{\mathbb{Z}}$-expectation in (3.22) is less than ϵ for all $n \geq c(\epsilon)$. Hence, for any N such that $c(\epsilon) \leq a_N$, we have $c(\epsilon) \leq a_N \leq D_{a_N}$ and the expression in (3.22) is less than ϵ. $\square$

4. Excursions are almost independent

The purpose of this section is to derive an estimate on the continuous-time excursions $(\mathsf{X}_{[R_k, D_k]})_{1 \leq k \leq k_*}$ between C and the complement of O. The main result is Lemma 4.3, showing that these excursions can essentially be replaced by independent excursions after conditioning on the $\mathbb{Z}$-projections of the successive return and departure points. The reason is that the G_N-component of X has enough time to mix and become close to uniformly distributed between every departure and subsequent return, thanks to the choice of h_N in the definition of the grids $\mathcal{G}_N$, see (3.4). The following estimate is the crucial ingredient:

PROPOSITION 4.1.

(4.1) $$\sup_{y, y' \in G_N} \left| q_t^{G_N}(y, y') - \frac{1}{|G_N|} \right| \leq e^{-\lambda_N t}, \text{ for } t \geq 0.$$

PROOF. If $w_y = 1$ for all $y \in G$, then the statement is immediate from [**28**], Corollary 2.1.5, page 328. As we now show, the argument given in [**28**] extends to the present context. For any $|G| \times |G|$ matrix A and real-valued function f on G, we define the function Af by

$$Af(y) = \sum_{y' \in G} A_{y,y'} f(y').$$

We define the matrices K and W by $K_{y,y'} = p^G(y, y')$ and $W_{y,y'} = w_y \delta_{y=y'}$, for $y, y' \in G$. Then we claim that for any real-valued function f on G,

(4.2) $E_y[f(\mathsf{Y}_t)] = H_t f(y)$, where $H_t = e^{-tW(I-K)}$, $t \geq 0$.

In words, this claim asserts that the infinitesimal generator matrix Q of the Markov chain $(\mathsf{Y}_t)_{t \geq 0}$ is given by $Q = -W(I - K)$, an elementary fact that is proved in [**23**], Theorem 2.8.2, p. 94. Recall the definition of the Dirichlet form $\mathcal{D}$ from (2.8). Let us also define the inner product of real-valued functions f and g on G by

$$\langle f, g \rangle = \sum_{y \in G} f(y) g(y) |G|^{-1}.$$

4. Excursions are almost independent

Then elementary computations show that

$$\frac{d}{dt}\mu((H_t f)^2) = -2\langle W(I-K)H_t f, H_t f\rangle = -2\mathcal{D}(H_t f, H_t f).$$

This equation implies that the function u, defined by $u(t) = \mathrm{var}_\mu(H_t f)$, $t \geq 0$, satisfies

$$u'(t) = -2\mathcal{D}(H_t(f-\mu(f)), H_t(f-\mu(f))) \stackrel{(2.9)}{\leq} -2\lambda_N u(t), \, t \geq 0,$$

hence by integration of of u'/u,

(4.3) $\quad \mathrm{var}_\mu(H_t f) = u(t) \leq e^{-2\lambda_N t}u(0) = e^{-2\lambda_N t}\mathrm{var}_\mu(f).$

Using symmetry of $q_t^G(.,.)$, (4.2) and the Cauchy-Schwarz inequality for the first estimate, we obtain for any $t \geq 0$ and $y, y' \in G$,

$$||G|q_t^G(y,y') - 1| = \left|\sum_{y'' \in G}\left(|G|q_{t/2}^G(y,y'') - 1\right)\left(|G|q_{t/2}^G(y'',y') - 1\right)\frac{1}{|G|}\right|$$

$$\leq \mathrm{var}_\mu\left(H_{t/2}|G|\delta_y(.)\right)^{1/2}\mathrm{var}_\mu\left(H_{t/2}|G|\delta_{y'}(.)\right)^{1/2}$$

$$\stackrel{(4.3)}{\leq} e^{-\lambda_N t}\mathrm{var}_\mu\left(|G|\delta_y(.)\right)^{1/2}\mathrm{var}_\mu\left(|G|\delta_{y'}(.)\right)^{1/2}$$

$$= e^{-\lambda_N t}(|G| - 1).$$

Dividing both sides by $|G|$, we obtain (4.1). □

Next, we show that the time between any departure and successive return indeed is typically much longer than the relaxation time λ_N^{-1} of Y:

LEMMA 4.2.

(4.4) $\quad \limsup_N |G_N|^{-\epsilon/16} \log \sup_{k \geq 2} P_0^\mathsf{Z}[\mathsf{R}_k - \mathsf{D}_{k-1} \leq \lambda_N^{-1}|G_N|^\epsilon] < 0.$

PROOF. By (3.4), we may assume that N is large enough so that $d_N < h_N/2$. We put

$$\gamma = 2\lambda_N^{-1}|G_N|^{\epsilon/8},$$

so that γ diverges as N tends to infinity (see below A1), and define the stopping times $(\mathsf{U}_n)_{n \geq 1}$ as the times of successive displacements of Z at distance $[\sqrt{\gamma}]$, i.e.

$$\mathsf{U}_1 = \inf\{t \geq 0 : |\mathsf{Z}_t - \mathsf{Z}_0| \geq [\sqrt{\gamma}]\}, \text{ and for } n \geq 2,$$
$$\mathsf{U}_n = \mathsf{U}_1 \circ \theta_{\mathsf{U}_{n-1}} + \mathsf{U}_{n-1}.$$

To get from a point in O^c to C, Z has to travel a distance of at least $h_N/2 \geq [h_N/(2\sqrt{\gamma})][\sqrt{\gamma}]$. As a consequence, $\mathsf{R}_k - \mathsf{D}_{k-1} \geq \mathsf{U}_{[h_N/(2\sqrt{\gamma})]} \circ \theta_{\mathsf{D}_{k-1}}$ and it follows from the strong Markov property applied at time D_{k-1}, then inductively at the times $\mathsf{U}_{[h_N/(2\sqrt{\gamma})]-1}, \ldots, \mathsf{U}_1$ that

(4.5) $\quad P_0^\mathsf{Z}[\mathsf{R}_k - \mathsf{D}_{k-1} \leq \gamma] \leq eE_0^\mathsf{Z}[\exp\{-\mathsf{U}_{[h_N/(2\sqrt{\gamma})]}/\gamma\}]$

$$\leq e\bigl(E_0^\mathsf{Z}[\exp\{-\mathsf{U}_1/\gamma\}]\bigr)^{[h_N/(2\sqrt{\gamma})]}.$$

124   4. Random walks on cylinders and random interlacements

Since $\mathsf{U}_1 = \mathsf{T}_{(-[\sqrt{\gamma}],[\sqrt{\gamma}])} = \sigma^{\mathbb{Z}}_{T_{(-[\sqrt{\gamma}],[\sqrt{\gamma}])}}$, we find with independence of $(\sigma^{\mathbb{Z}}_n)_{n\geq 0}$ and $T_{(-[\sqrt{\gamma}],[\sqrt{\gamma}])}$,

$$E^{\mathbb{Z}}_0[\exp\{-\mathsf{U}_1/\gamma\}] = E^{\mathbb{Z}}_0\left[(1-1/\gamma)^{T_{(-[\sqrt{\gamma}],[\sqrt{\gamma}])}}\right],$$

by computing the moment generating function of the $\Gamma(n,1)$-distributed variable $\sigma^{\mathbb{Z}}_n$. By the invariance principle, the last expectation is bounded from above by $1-c$ for some constant $c>0$. Inserting this bound into (4.5) and using the bound $h_N \geq c\sqrt{\gamma}|G_N|^{\epsilon/16}$ from (3.4), we find (4.4). □

We finally come to the announced result, which is similar to Proposition 3.3 in [36]. We introduce, for $\mathcal{G}$ any one of the graphs G_N, $\mathbb{Z}$ or $G_N \times \mathbb{Z}$, the spaces $\mathcal{P}(\mathcal{G})^f$ of right-continuous functions from $[0,\infty)$ to $\mathcal{G}$ with finitely many discontinuities, endowed with the canonical σ-algebras generated by the finite-dimensional projections. The measurable functions $(.)^{s_1}_{s_0}$ from $\mathcal{P}(\mathcal{G})$ to $\mathcal{P}(\mathcal{G})^f$ are defined for $0 \leq s_0 < s_1$ by

(4.6) $((\mathsf{w})^{s_1}_{s_0})_t = \mathsf{w}_{(s_0+t)\wedge s_1}$, $t \geq 0$.

Given $z \in C$ and z' with $P_z[\mathsf{Z}_{\mathsf{D}_1} = z'] > 0$, for P_z defined in (2.4) (in other words $z' \in \partial \tilde{I}$ if $\partial \tilde{I}$ is the connected component of O containing z), we set

(4.7) $P_{z,z'} = P_z[.|\mathsf{Z}_{\mathsf{D}_1} = z']$.

LEMMA 4.3. *For any measurable functions* $f_k : \mathcal{P}(G_N)^f \times \mathcal{P}(\mathbb{Z})^f \to [0,1]$, $1 \leq k \leq k_*$,

(4.8) $\lim_N \left| E\left[\prod_{1\leq k \leq k_*} f_k((\mathsf{X})^{\mathsf{D}_k}_{\mathsf{R}_k})\right] - E^{\mathbb{Z}}_0\left[\prod_{1\leq k \leq k_*} E_{\mathsf{Z}_{\mathsf{R}_k},\mathsf{Z}_{\mathsf{D}_k}}[f_k((\mathsf{X})^{\mathsf{D}_1}_0)]\right]\right| = 0.$

PROOF OF LEMMA 4.3. Consider first arbitrary measurable functions $g_k : \mathcal{P}(G)^f \to [0,1]$, $1 \leq k \leq k_*$, real numbers $0 \leq s_1 < s'_1 < \ldots < s_{k_*} < s'_{k_*} < \infty$ and set

$$H_k = g_k((\mathsf{Y})^{s'_k}_{s_k}).$$

With the simple Markov property applied at time s_{k_*}, then at time s_{k_*-1}, one obtains

$$E^G\left[\prod_{1\leq k\leq k_*} H_k\right] = E^G\left[\left(\prod_{1\leq k\leq k_*-1} H_k\right) E^G_{\mathsf{Y}_{s_{k_*}}}[g_{k_*}((\mathsf{Y})^{s'_{k_*}-s_{k_*}}_0)]\right]$$

$$= E^G\left[\left(\prod_{1\leq k\leq k_*-1} H_k\right) \sum_{y\in G} q^G_{s_{k_*}-s'_{k_*-1}}(\mathsf{Y}_{s_{k_*-1}}, y)\right] E^G_y[g_{k_*}((\mathsf{Y})^{s'_{k_*}-s_{k_*}}_0)].$$

With the estimate (4.1) on the difference between the transition probability of Y inside the expectation and the uniform distribution and the

4. Excursions are almost independent   125

fact that $g_k \in [0,1]$, it follows that

$$\left| E^G\Big[\prod_{1\leq k\leq k_*} H_k\Big] - E^G\Big[\prod_{1\leq k\leq k_*-1} H_k\Big] E^G[g_{k_*}((Y)_0^{s'_{k_*}-s_{k_*}})] \right|$$
$$\leq c|G|\exp\{-(s_{k_*}-s'_{k_*-1})\lambda_N\}.$$

By induction, we infer that

(4.9) $$\left| E^G\Big[\prod_{1\leq k\leq k_*} g_k((Y)_{s_k}^{s'_k})\Big] - \prod_{1\leq k\leq k_*} E^G[g_k((Y)_0^{s'_k-s_k})] \right|$$
$$\leq c|G| \sum_{2\leq k\leq k_*} e^{-(s_k-s'_{k-1})\lambda_N}.$$

Let us now consider the first expectation in (4.8). By Fubini's theorem, we find that

$$E\Big[\prod_{1\leq k\leq k_*} f_k((X)_{\mathsf{R}_k}^{\mathsf{D}_k})\Big] = E_0^{\mathbb{Z}}\Big[E^G\Big[\prod_{1\leq k\leq k_*} f_k((Y)_{s_k}^{s'_k},(\bar{z})_{s_k}^{s'_k})\Big]\Big|_{(\bar{z})_{s_k}^{s'_k}=(Z)_{\mathsf{R}_k}^{\mathsf{D}_k}}\Big].$$

Observe that (4.9) applies to the E^G-expectation with

$$g_k(.) = f_k(.,(\bar{z})_{s_k}^{s'_k}),$$

and yields

(4.10) $$\left| E\Big[\prod_{1\leq k\leq k_*} f_k((X)_{\mathsf{R}_k}^{\mathsf{D}_k})\Big] - E_0^{\mathbb{Z}}\Big[\prod_{1\leq k\leq k_*} E^G\big[f_k((Y)_0^{s'_k-s_k},(Z)_{\mathsf{R}_k}^{\mathsf{D}_k})\big]\Big] \right|$$
$$\leq c|G| \sum_{2\leq k\leq k_*} E_0^{\mathbb{Z}}[e^{-(\mathsf{R}_k-\mathsf{D}_{k-1})\lambda_N}].$$

Note that for large N, the last term can be bounded with the estimate (4.4) on $\mathsf{R}_k - \mathsf{D}_{k-1}$:

(4.11) $$\sum_{2\leq k\leq k_*} E_0^{\mathbb{Z}}[e^{-(\mathsf{R}_k-\mathsf{D}_{k-1})\lambda_N}] \leq ck_* \exp\{-c'|G|^{c\varepsilon}\}$$
$$\stackrel{(3.10)}{\leq} c(\alpha)|G|^c \exp\{-c'|G|^{c\varepsilon}\}.$$

It thus only remains to show that the second expectation on the left-hand side of (4.10) is equal to the second expectation in (4.8). Note that for any measurable functions $h_k : \mathcal{P}(\mathbb{Z})^f \to [0,1]$, $1 \leq k \leq k_*$ and points $z_1,\ldots,z_{k_*}$, $z'_1,\ldots,z'_{k_*}$ in $\mathbb{Z}$ such that $P_{z_k}^{\mathbb{Z}}[Z_{\mathsf{D}_1} = z'_k] > 0$ for $1 \leq k \leq k_*$, one has by two successive inductive applications of the strong Markov property at the times $\mathsf{R}_{k_*}, \mathsf{D}_{k_*-1}, \mathsf{R}_{k_*-1},\ldots,\mathsf{D}_1$, with the convention $P_{z'_0} = P$,

$$E_0^{\mathbb{Z}}\Big[\bigcap_{1\leq k\leq k_*}\{\mathsf{Z}_{\mathsf{R}_k} = z_k, \mathsf{Z}_{\mathsf{D}_k} = z'_k\}, \prod_{1\leq k\leq k_*} h_k((\mathsf{Z})_{\mathsf{R}_k}^{\mathsf{D}_k})\Big]$$

$$= \prod_{1\leq k\leq k_*}\Big(P_{z'_{k-1}}^{\mathbb{Z}}[\mathsf{Z}_{\mathsf{R}_1} = z_k] E_{z_k,z'_k}[h_k((\mathsf{Z})_0^{\mathsf{D}_1})] P_{z_k}^{\mathbb{Z}}[\mathsf{Z}_{\mathsf{D}_1} = z'_k]\Big)$$

$$= P_0^{\mathbb{Z}}\Big[\bigcap_{1\leq k\leq k_*}\{\mathsf{Z}_{\mathsf{R}_k} = z_k, \mathsf{Z}_{\mathsf{D}_k} = z'_k\}\Big]\prod_{1\leq k\leq k_*} E_{z_k,z'_k}[h_k((\mathsf{Z})_0^{\mathsf{D}_1})].$$

Summing this last equation over all z_k, z'_k as above, one obtains

$$E_0^{\mathbb{Z}}\Big[\prod_{1\leq k\leq k_*} h_k((\mathsf{Z})_{\mathsf{R}_k}^{\mathsf{D}_k})\Big] = E_0^{\mathbb{Z}}\Big[\prod_{1\leq k\leq k_*} E_{\mathsf{Z}_{\mathsf{R}_k},\mathsf{Z}_{\mathsf{D}_k}}[h_k((\mathsf{Z})_0^{\mathsf{D}_1})]\Big].$$

Applying this equation with

$$h_k((\mathsf{Z})_{\mathsf{R}_k}^{\mathsf{D}_k}) = E^G\big[f_k((\mathsf{Y})_0^{s'_k-s_k}, (\tilde{z})_{s_k}^{s'_k})\big]\Big|_{(\tilde{z})_{s.}^{s'_{\cdot}}=(\mathsf{Z})_{\mathsf{R}_k}^{\mathsf{D}_k}},$$

substituting the result into (4.10) and remembering (4.11), we have shown (4.8). $\square$

5. Proof of the result in continuous time

The purpose of this section is to prove in Theorem 5.1 the continuous-time version of Theorem 1.1. Let us explain the role of the crucial estimates appearing in Lemmas 5.2 and 5.3. Under the assumptions A1-A10, these lemmas exhibit the asymptotic behavior of the $P_{z,z'}$-probability (see (4.7)) that an excursion of the path X visits vertices in the neighborhoods of the sites x_m contained in a box $G_N \times I$. It is in particular shown that the probability that a set V_m in the neighborhood of x_m is visited equals $\mathrm{cap}^m(\Phi_m(V_m))h_N/|G_N|$, up to a multiplicative factor tending to 1 as N tends to infinity. This estimate is similar to a more precise result proved by Sznitman for $G_N = (\mathbb{Z}/N\mathbb{Z})^d$ in Lemma 1.1 of [37], where an identity is obtained for the same probability, if the distribution of the starting point of the excursion is the uniform distribution on the boundary of $G_N \times \tilde{I}$ (rather than the uniform distribution on $G_N \times \{z\}$).

According to the characterization (1.5), these crucial estimates show that the law of the vertices in the neighborhood of x_m not visited by such an excursion is comparable to $\mathbb{Q}_{h_N/|G_N|}^{G_m\times\mathbb{Z}}$. In Lemma 4.3 of the previous section, we have seen that different excursions of the form $(\mathsf{X})_{\mathsf{R}_k}^{\mathsf{D}_k}$, conditioned on the entrance and departure points of the $\mathbb{Z}$-projection, are close to independent for large N. According to the observation outlined in the last paragraph, the level of the random interlacement appearing in the neighborhood of x_m at time $\alpha|G_N|^2$ is hence approximately equal to $h_N/|G_N|$ times the number of excursions to the interval I performed until time $\alpha|G_N|^2$. As we have seen in Proposition 3.3

5. Proof of the result in continuous time

and Lemma 3.4, this quantity is close to the local time $\hat{L}^{z_m}_{\alpha|G_N|^2}/|G_N|$ for large N. An invariance principle for local times due to Révész [**25**] (with assumption A4) serves to identify the limit of this quantity, hence the level of the random interlacement appearing in the large N limit, as $L(v_m, \alpha)$. This strategy will yield the following result:

THEOREM 5.1. *Assume that* A1-A10 *are satisfied. Then the graphs* $\mathbb{G}_m \times \mathbb{Z}$ *are transient and as N tends to infinity, the* $\prod_{m=1}^{M} \{0,1\}^{\mathbb{G}_m} \times \mathbb{R}_+^M$-*valued random variables*

$$\left(\omega^{1,N}_{\eta^{\times}_{\alpha|G_N|^2}}, \ldots, \omega^{M,N}_{\eta^{\times}_{\alpha|G_N|^2}}, \frac{\mathsf{L}^{z_1}_{\alpha|G_N|^2}}{|G_N|}, \ldots, \frac{\mathsf{L}^{z_M}_{\alpha|G_N|^2}}{|G_N|} \right), \, \alpha > 0,$$

defined by (1.4), (2.6), with r_N *and* $\phi_{m,N}$ *chosen in (5.1) and (5.2), converge in joint distribution under P to the law of the random vector*

$$(\omega_1, \ldots, \omega_M, U_1, \ldots, U_m)$$

with the following distribution: $(U_m)_{m=1}^{M}$ *is distributed as*

$$(L(v_m, \alpha))_{m=1}^{M}$$

under W, and conditionally on $(U_m)_{m=1}^{M}$, *the random variables* $(\omega_m)_{m=1}^{M}$ *have joint distribution*

$$\prod_{1 \leq m \leq M} \mathbb{Q}_{U_m}^{\mathbb{G}_m \times \mathbb{Z}}.$$

PROOF. The transience of the graphs $\mathbb{G}_m \times \mathbb{Z}$ is an immediate consequence of Lemma 2.3. To define the local pictures in (1.4), we choose the r_N in (1.3) as

(5.1) $\qquad \mathsf{r}_N = \left(\min_{1 \leq m < m' \leq M} d(x_{m,N}, x_{m',N}) \wedge r_N \wedge d_N \right)/3,$

cf. A3, A5, (3.4)

(5.2) $\qquad$ and $\phi_{m,N}$ as the restriction of the isomorphism in A5 to $B(y_{m,N}, \mathsf{r}_N)$.

Then the local pictures in (1.4) are defined. We set

(5.3) $\quad B_{m,N} = B(x_{m,N}, \mathsf{r}_N - 1)$ and $\mathbb{B}_{m,N} = \Phi_{m,N}(B_{m,N})$, for $\mathsf{r}_N \geq 1$.

From now on, we drop N from the notation in $\phi_{m,N}$, $B_{m,N}$ and $\mathbb{B}_{m,N}$ for simplicity. Our present task is to show that for arbitrarily chosen finite subsets $\mathbb{V}_m$ of $\mathbb{G}_m \times \mathbb{Z}$,

(5.4) $\quad A_N(\alpha|G_N|^2, \alpha|G_N|^2) \to A(\alpha)$, for any $\theta_m \in \mathbb{R}_+$, $1 \leq m \leq M$.

128   4. Random walks on cylinders and random interlacements

where for times $s, s' \geq 0$ and $V_m = \Phi_m^{-1}\mathbb{V}_m$ (well-defined for large N, see (2.10)),

(5.5) $\quad A_N(s, s') = E\Big[\prod_{1\leq m \leq M} \mathbf{1}_{\{H_{V_m} > s\}} \exp\Big\{-\frac{\theta_m}{|G_N|} L_{s'}^{z_m}\Big\}\Big]$, and

(5.6) $\quad A(\alpha) = E^W\Big[\exp\Big\{-\sum_{1\leq m \leq M} L(v_m, \alpha)(\mathrm{cap}^m(\mathbb{V}_m) + \theta_m)\Big\}\Big]$.

Theorem 5.1 then follows, as a result of the equivalence of weak convergence and convergence of Laplace transforms (see for example [**7**], p. 189-191), the compactness of the set of probability measures on $\prod_m \{0,1\}^{\mathbb{G}_m \times \mathbb{Z}}$, and the fact that the canonical product σ-algebra on $\prod_m \{0,1\}^{\mathbb{G}_m \times \mathbb{Z}}$ is generated by the π-system of events $\cap_{m=1}^M \{\omega(x) = 1, \text{ for all } x \in \mathbb{V}_m\}$, with $\mathbb{V}_m$ varying over finite subsets of $\mathbb{G}_m \times \mathbb{Z}$.

We first introduce some additional notation and state some inclusions we shall use. For any interval $I \in \mathcal{I}$ (cf. (3.2)), we denote by $\mathcal{J}_I$ the set of indices m such that $z_m \in I$:

(5.7) $\quad \mathcal{J}_I = \{1 \leq m \leq M : z_{m,N} \in I\}/ = \emptyset$ if no $z_{m,N}$ belongs to I.

Note that the set $\mathcal{J}_I$ depends on N. Indeed, so does the labelling of the intervals I_l in $\mathcal{I}$. It follows from the definition of r_N that

(5.8) $\quad$ the balls $(\bar{B}_m)_{1 \leq m \leq M}$ are disjoint, cf. (5.3).

Since the sets $\mathbb{V}_m$ are finite, we can choose a parameter $\kappa > 0$ such that $\mathbb{V}_m \subset B((o_m, 0), \kappa)$ for all m and N. Since r_N tends to infinity with N, there is an $N_0 \in \mathbb{N}$ such that for all $N \geq N_0$, we have $\mathsf{r}_N \geq 1$ as well as for all $I \in \mathcal{I}$ and $m \in \mathcal{J}_I$,

(5.9)
$$\begin{array}{ccccccc} \mathbb{V}_m & \subset & B((o_m, 0), \kappa) & \subset & \mathbb{B}_m & \subset & B(o_m, \mathsf{r}_N - 1) \times \mathbb{Z} \\ \downarrow \Phi_m^{-1} & & \downarrow \Phi_m^{-1} & & \downarrow \Phi_m^{-1} & & \\ V_m & \subset & B(x_m, \kappa) & \subset & B_m & \overset{(5.1)}{\subset} & B(y_m, \mathsf{r}_N - 1) \times I. \end{array}$$

Since $d_N = o(|G_N|)$ (cf. (3.4), A2), any two sequences z_m that are contained in the same interval $I \in \mathcal{I}$ infinitely often, when divided by $|G_N|$, must converge to the same number v_m, cf. (A4). By A8, we can hence increase N_0 if necessary, such that for all $N \geq N_0$,

(5.10) $\quad$ for m and m' in $\mathcal{J}_I$, either $C_m = C_{m'}$ or $C_m \cap C_{m'} = \emptyset$.

We use $V_{I,m}$ to denote the union of all sets $V_{m'}$ included in $C_m \times I$ and V_I for the union of all V_m included in $G_N \times I$, i.e.

(5.11) $\quad V_{I,m} = \bigcup_{m' \in \mathcal{J}_I : C_{m'} = C_m} V_{m'} \subset C_m \times I$, and

$$V_I = \bigcup_{m \in \mathcal{J}_I} V_m \overset{(5.9)}{\subset} G_N \times I,$$

5. Proof of the result in continuous time

with the convention that the union of no sets is the empty set.

The proof of (5.4) uses three additional Lemmas that we now state. The first two lemmas show that the probability that the continuous-time random walk X started from the boundary of $G_N \times I$ hits a point in the set $V_I \subset G_N \times I$ (cf. (5.11)) before exiting $G \times \tilde{I}$ behaves like $h_N/|G_N|$ times the sum of the capacities of those sets $\mathbb{V}_m$ whose preimages under Φ_m are subsets of $G_N \times I$.

LEMMA 5.2. *Under A1-A10, for $N \geq N_0$ (cf. (5.9), (5.10)), any $I \in \mathcal{I}$, $I \subset \tilde{I} \in \tilde{\mathcal{I}}$, $z_1 \in \partial(I^c)$ and $z_2 \in \partial \tilde{I}$,*

$$(5.12) \quad 1 - c\frac{d_N}{h_N} \leq P_{z_1,z_2}[\mathsf{H}_{V_I} < \mathsf{T}_{\tilde{B}}]\left(\frac{h_N}{|G_N|} \operatorname{cap}_{\tilde{B}}(V_I)\right)^{-1} \leq 1 + c\frac{d_N}{h_N},$$

where $\tilde{B} = G_N \times \tilde{I}$ and $\operatorname{cap}_{\tilde{B}}(V_I) = \sum_{x \in V_I} P_x[T_{\tilde{B}} < \tilde{H}_{V_I}]w_x$.

LEMMA 5.3. *With the assumptions and notation of Lemma 5.2,*

$$(5.13) \quad \lim_N \max_{I \in \mathcal{I}} \left| \operatorname{cap}_{\tilde{B}}(V_I) - \sum_{m \in \mathcal{J}_I} \operatorname{cap}^m(\mathbb{V}_m) \right| = 0.$$

The next lemma allows to disregard the the effect of the random walk trajectory until time D_1, cf. (5.15), as well as the difference between D_{k^*} and D_{k_*}, cf. (5.16).

LEMMA 5.4. *Assuming A1,*

$$(5.14) \quad \lim_N \sup_{z \in \mathbb{Z}, x \in G_N \times \mathbb{Z}} P_z[\mathsf{H}_x \leq \mathsf{D}_{k^*-k_*}] = 0.$$

$$(5.15) \quad \lim_N \sup_{z \in \mathbb{Z}} P_z[\mathsf{H}_{\cup_I V_I} \leq \mathsf{D}_1] = 0.$$

$$(5.16) \quad \lim_N E\left[\left| \prod_{1 \leq m \leq M} 1_{\{\mathsf{H}_{\mathbb{V}_m} > \mathsf{D}_{k^*}\}} - \prod_{1 \leq m \leq M} 1_{\{\mathsf{H}_{\mathbb{V}_m} > \mathsf{D}_{k_*}\}} \right|\right] = 0.$$

Before we prove Lemmas 5.2-5.4, we show that they allow us to deduce Theorem 5.1. Throughout the proof, we set $T = \alpha|G_N|^2$ and say that two sequences of real numbers are limit equivalent if their difference tends to 0 as N tends to infinity. We first claim that in order to show (5.4), it is sufficient to prove that

$$(5.17) \quad A'_N = A_N(\mathsf{D}_{k_*}, T) \to A(\alpha), \text{ for } \alpha > 0.$$

Indeed, by (5.16), the statement (5.17) implies that also

$$(5.18) \quad \lim_N A_N(\mathsf{D}_{k^*}, T) = A(\alpha), \text{ for } \alpha > 0.$$

Now recall that $\mathsf{D}_{k_*} \leq T \leq \mathsf{D}_{k^*}$ with probability tending to 1 by (3.11). Together with (3.21), it follows that

$$\lim_N P_0^{\mathbb{Z}}[(1-\delta)\mathsf{D}_{k_*} \leq T \leq (1+\delta)\mathsf{D}_{k^*}] = 1, \text{ for any } \delta > 0.$$

130   4. Random walks on cylinders and random interlacements

Monotonicity in both arguments of $A_N(.,.)$, (5.17) and (5.18) hence yield

$$\limsup_N A_N\bigl(T/(1-\delta),T/(1-\delta)\bigr) \le \limsup_N A_N(\mathsf{D}_{k_*},T) = A(\alpha) \text{ and}$$

$$\liminf_N A_N\bigl(T/(1+\delta),T/(1+\delta)\bigr) \ge \liminf_N A_N(\mathsf{D}_{k^*},T) = A(\alpha),$$

for $0 < \delta < 1$. Replacing α by $\alpha(1-\delta)$ and $\alpha(1+\delta)$ respectively, we deduce that

$$A(\alpha(1+\delta)) \le \liminf_N A_N(T,T) \le \limsup_N A_N(T,T) \le A(\alpha(1-\delta)),$$

for $\alpha > 0$ and $0 < \delta < 1$, from which (5.4) follows by letting δ tend to 0 and using the continuity of $A(.)$. Hence, it suffices to show (5.17). By (3.17) A'_N is limit equivalent to

$$(5.19) \qquad E\Bigl[\mathbf{1}_{\cap_m\{\mathsf{H}_{V_m} > \mathsf{D}_{k_*}\}} \exp\Bigl\{-\sum_{1 \le m \le M} \frac{\theta_m}{|G_N|} \hat{L}_{[T]}^{z_m}\Bigr\}\Bigr],$$

which by (5.15) remains limit equivalent if the event $\cap_m\{\mathsf{H}_{V_m} > \mathsf{D}_{k_*}\}$ is replaced by

$$\mathcal{A} = \Bigl\{\text{for all } 2 \le k \le k_*, \text{ if } \mathsf{Z}_{\mathsf{R}_k} \in I \text{ for some } I \in \mathcal{I}, \mathsf{X}_{[\mathsf{R}_k,\mathsf{D}_k]} \cap V_I = \emptyset\Bigr\},$$

cf. (3.2). Making use of (3.12) and (3.14) (together with $Z_{R_k} = \mathsf{Z}_{\mathsf{R}_k}$) we find that A'_N is limit equivalent to (cf. (5.7))

$$(5.20) \qquad E\Bigl[\mathbf{1}_\mathcal{A} \exp\Bigl\{-\sum_{1 \le l \le M} \frac{h_N(\sum_{m \in \mathcal{J}_{I_l}} \theta_m)}{|G_N|} \sum_{1 \le k \le k_*} \mathbf{1}_{\{\mathsf{Z}_{\mathsf{R}_k} \in I_l\}}\Bigr\}\Bigr].$$

Since $h_N = o(|G_N|)$ (cf. (3.4), A2), this expectation remains limit equivalent if we drop the $k = 1$ term in the second sum. In other words, the expression in (5.20) is limit equivalent to (recall the notation from (4.6))

$$E\Bigl[\prod_{k=2}^{k_*} f((\mathsf{X})_{\mathsf{R}_k}^{\mathsf{D}_k})\Bigr], \text{ with } f: \mathcal{P}(G_N)^f \times \mathcal{P}(\mathbb{Z})^f \to [0,1] \text{ defined by}$$

$$f(\mathsf{w}) = \prod_{1 \le l \le M} \Bigl\{\Bigl(1 - \mathbf{1}_{\{\mathsf{w}_0 \in G_N \times I_l\}} \mathbf{1}_{\{\mathsf{w}_{[0,\infty)} \cap V_{I_l} \ne \emptyset\}}\Bigr)$$

$$\exp\Bigl(-\frac{h_N(\sum_{m \in \mathcal{J}_{I_l}} \theta_m)}{|G_N|} \mathbf{1}_{\{\mathsf{w}_0 \in G_N \times I_l\}}\Bigr)\Bigr\}.$$

By Lemma 4.3 with $f_1 = 1$, $f_k = f$ for $2 \le k \le k_*$, A'_N is hence limit equivalent to

$$E_0^\mathbb{Z}\Bigl[\prod_{2 \le k \le k_*} E_{\mathsf{Z}_{\mathsf{R}_k},\mathsf{Z}_{\mathsf{D}_k}}[f((\mathsf{X})_0^{\mathsf{D}_1})]\Bigr].$$

5. Proof of the result in continuous time

The above expression equals

$$(5.21) \qquad E_0^{\mathbb{Z}}\Bigg[\prod_{\substack{2\leq k\leq k_*\\1\leq l\leq M}}\bigg\{\Big(1-\mathbf{1}_{\{Z_{R_k}\in I_l\}}g_l(Z_{R_k},Z_{D_k})\Big)$$
$$\exp\Big\{-\frac{h_N(\sum_{m\in\mathcal{J}_{I_l}}\theta_m)}{|G_N|}\mathbf{1}_{\{Z_{R_k}\in I_l\}}\Big\}\bigg\}\Bigg],$$

where $g_l(z,z') = P_{z,z'}\big[X_{[0,D_1]}\cap V_{I_l}\neq\emptyset\big]$.

From (5.12), we know that

$$(5.22) \qquad 1-c\frac{d_N}{h_N} \leq g_l(Z_{R_k},Z_{D_k})\Big(\frac{h_N}{|G_N|}\mathrm{cap}_{G\times\tilde{I}_l}(V_I)\Big)^{-1} \leq 1+c\frac{d_N}{h_N}.$$

With the inequality $0\leq e^{-u}-1+u\leq u^2$ for $u\geq 0$, one obtains that

$$\Big|\prod_{\substack{2\leq k\leq k_*\\1\leq l\leq M}}\big(1-\mathbf{1}_{\{Z_{R_k}\in I_l\}}g\big) - \prod_{\substack{2\leq k\leq k_*\\1\leq l\leq M}}\exp\big\{-\mathbf{1}_{\{Z_{R_k}\in I_l\}}g\big\}\Big|$$
$$\leq \sum_{\substack{2\leq k\leq k_*\\1\leq l\leq M}}\mathbf{1}_{\{Z_{R_k}\in I_l\}}g^2,$$

where we have witten g in place of $g_l(Z_{R_k},Z_{D_k})$. The expectation of the right-hand side in the last estimate tends to 0 as N tends to infinity, thanks to (5.22) and (3.13). The expression in (5.21) thus remains limit equivalent to A'_N if we replace $1-\mathbf{1}_{\{Z_{R_k}\in I_l\}}g_l(Z_{R_k},Z_{D_k})$ by $\exp\{-\mathbf{1}_{\{Z_{R_k}\in I_l\}}g_l(Z_{R_k},Z_{D_k})\}$. Using again (3.13), together with (5.13) and (5.22), we may then replace $g_l(Z_{R_k},Z_{D_k})$ by $\frac{h_N}{|G_N|}\sum_{m\in\mathcal{J}_{I_l}}\mathrm{cap}^m(\mathbb{V}_m)$. We deduce that the following expression is limit equivalent to A'_N:

$$E_0^{\mathbb{Z}}\Big[\exp\big\{-\sum_{\substack{1\leq k\leq k_*\\1\leq l\leq M}}\sum_{m\in\mathcal{J}_{I_l}}\frac{h_N}{|G_N|}\mathbf{1}_{\{Z_{R_k}\in I_l\}}\big(\mathrm{cap}^m(\mathbb{V}_m)+\theta_m\big)\big\}\Big].$$

By (3.14) and (3.12), this expression is also limit equivalent to

$$(5.23) \qquad E_0^{\mathbb{Z}}\Big[\exp\big\{-\sum_{1\leq m\leq M}\frac{1}{|G_N|}\hat{L}^{zm}_{[T]}\big(\mathrm{cap}^m(\mathbb{V}_m)+\theta_m\big)\big\}\Big].$$

With Proposition 1 in [**25**], one can construct a coupling of the simple random walk Z on $\mathbb{Z}$ with a Brownian motion on $\mathbb{R}$ such that for any $\rho>0$,

$$n^{-1/4-\rho}\sup_{z\in\mathbb{Z}}\big|\hat{L}^z_n - L(z,n)\big| \xrightarrow{n\to\infty} 0, \quad a.s.,$$

where $L(.,.)$ is a jointly continuous version of the local time of the canonical Brownian motion. It follows that (5.23), hence A'_N is limit

132   4. Random walks on cylinders and random interlacements

equivalent to

(5.24) $\quad E^W\Big[\exp\Big\{-\sum_{1\le m\le M}\frac{1}{|G_N|}L(z_m,[\alpha|G_N|^2])\big(\mathrm{cap}^m(\mathbb{V}_m)+\theta_m\big)\Big\}\Big].$

By Brownian scaling, $L(z_m,[\alpha|G|^2])/|G|$ has the same distribution as $L(z_m/|G|,[\alpha|G|^2]/|G|^2)$.

Hence, the expression in (5.24) converges to $A(\alpha)$ in (5.6) by continuity of L and convergence of $z_m/|G|$ to v_m, see A4. We have thus shown that $A'_N \to A(\alpha)$ and by (5.17) completed the proof of Theorem 5.1. $\square$

We still have to prove Lemmas 5.2-5.4. To this end, we first show that the random walk X started at $\partial C_m \times I$ typically escapes from $G_N \times \tilde{I}$ before reaching a point in the vicinity of x_m. Here, the upper bound on h_N in (3.4) plays a crucial role.

LEMMA 5.5. *Assuming A1-A10, for any fixed vertex* $\mathbf{x} = (\mathbf{y}, z) \in \mathbb{G}_m \times \mathbb{Z}$, *intervals* $I \in \mathcal{I}$, $I \subset \tilde{I} \in \tilde{\mathcal{I}}$ *(cf. (3.2)) and* $z_m \in I$,

(5.25) $\quad \lim_N \sup_{y_0 \in \partial(C_m^c), z_0 \in \mathbb{Z}} P_{(y_0,z_0)}[H_{\Phi_m^{-1}(\mathbf{x})} < T_{G_N \times \tilde{I}}] = 0.$

(Note that $\Phi_m^{-1}(\mathbf{x})$ *is well-defined for large* N *by A5.)*

PROOF OF LEMMA 5.5. Consider any $x_0 = (y_0, z_0)$ with $y_0 \in \partial(C_m^c)$ and $z_0 \in \mathbb{Z}$. In order to bound the expectation of $T_{G\times\tilde{I}}$, recall that $T_{\tilde{I}}$ denotes the exit time of the interval $\tilde{I}$ by the discrete-time process Z, so that $T_{G\times\tilde{I}}$ can be expressed as $T_{\tilde{I}}$ plus the number of jumps Y makes until $T_{\tilde{I}}$. Since Y and Z, hence η^Y and σ^Z, are independent under P_{x_0}, this implies with Fubini's theorem and stochastic domination of η^Y by the Poisson process η^{c_1} (cf. (2.16)) that

$$E_{x_0}[T_{G\times\tilde{I}}] = E_{z_0}^{\mathbb{Z}}\big[T_{\tilde{I}} + E_{y_0}^G[\eta^Y_{\sigma^Z_{T_{\tilde{I}}}}]\big]$$
$$\le E_{z_0}^{\mathbb{Z}}[T_{\tilde{I}}] + c_1 E_{z_0}^{\mathbb{Z}}[\sigma^Z_{T_{\tilde{I}}}]$$
$$= (1+c_1)E_{z_0}^{\mathbb{Z}}[T_{\tilde{I}}] \le ch_N^2,$$

using a standard estimate on one-dimensional simple random walk in the last step. Hence by the Chebyshev inequality and the bound (3.4) on h_N,

$$P_{x_0}[T_{G\times\tilde{I}} \ge \lambda_N^{-1}|G|^\epsilon] \le E_{x_0}[T_{G\times\tilde{I}}]\lambda_N|G|^{-\epsilon} \le ch_N^2\lambda_N|G|^{-\epsilon} \le c|G|^{-\epsilon/2}.$$

The claim (5.25) thus follows from A10. $\square$

PROOF OF LEMMA 5.2. With z_1, z_2 as in the statement, we have by the strong Markov property applied at the hitting time of $V_I \subset G\times I$ (cf. (5.9)),

$$P_{z_1,z_2}[\mathsf{H}_{V_I} < \mathsf{T}_{\tilde{B}}] = P_{z_1}[\mathsf{H}_{V_I} < \mathsf{T}_{\tilde{B}}, \mathsf{Z}_{\mathsf{T}_{\tilde{B}}} = z_2]/P_{z_1}^{\mathbb{Z}}[\mathsf{Z}_{T_{\tilde{I}}} = z_2]$$
$$= E_{z_1}\big[\mathsf{H}_{V_I} < \mathsf{T}_{\tilde{B}}, P_{\mathsf{Z}_{\mathsf{H}_{V_I}}}^{\mathbb{Z}}[\mathsf{Z}_{\mathsf{T}_{\tilde{I}}} = z_2]\big]/P_{z_1}^{\mathbb{Z}}[\mathsf{Z}_{\mathsf{T}_{\tilde{I}}} = z_2].$$

5. Proof of the result in continuous time

From (3.4) and the definition of the intervals $I \subset \tilde{I}$, it follows that

$$\sup_{z \in I} \left| P_z^{\mathbb{Z}}[\mathsf{Z}_{\mathsf{T}_{\tilde{I}}} = z_2] - 1/2 \right| \leq c d_N/h_N,$$

hence from the previous equality that

(5.26) $\quad (1 - cd_N/h_N) P_{z_1}[\mathsf{H}_{V_I} < \mathsf{T}_{\tilde{B}}] \leq P_{z_1,z_2}[\mathsf{H}_{V_I} < \mathsf{T}_{\tilde{B}}]$, and
$$P_{z_1,z_2}[\mathsf{H}_{V_I} < \mathsf{T}_{\tilde{B}}] \leq P_{z_1}[\mathsf{H}_{V_I} < \mathsf{T}_{\tilde{B}}](1 + cd_N/h_N).$$

Note that $\{\mathsf{H}_{V_I} < \mathsf{T}_{\tilde{B}}\} = \{H_{V_I} < T_{\tilde{B}}\}$, P_{z_1}-a.s. Summing over all possible locations and times of the last visit of X to the set V_I, one thus finds

$$P_{z_1}[\mathsf{H}_{V_I} < \mathsf{T}_{\tilde{B}}] = \sum_{x \in V_I} \sum_{n=1}^{\infty} P_{z_1}\left[\{X_n = x, n < T_{\tilde{B}}\} \cap (\theta_n^X)^{-1}\{\tilde{H}_x > T_{\tilde{B}}\}\right].$$

After an application of the simple Markov property to the probability on the right-hand side, this last expression becomes

$$\sum_{x \in V_I} E_{z_1}\left[\sum_{n=1}^{T_{\tilde{B}}} \mathbf{1}_{\{X_n = x\}}\right] P_x[\tilde{H}_x > T_{\tilde{B}}]$$

$$= \sum_{x=(y,z) \in V_I} w_x E_{z_1}\left[\int_0^{\infty} \mathbf{1}_{\{\mathsf{Y}_t = y\}} \mathbf{1}_{\{\mathsf{Z}_t = z, t < \mathsf{T}_{\tilde{I}}\}} dt\right] P_x[\tilde{H}_x > T_{\tilde{B}}],$$

because the expected duration of each visit to x by X is $1/w_x$. Exploiting independence of Y and $(\mathsf{Z}, \mathsf{T}_{\tilde{I}})$ and the fact that Y_t is distributed according to the uniform distribution on G under P_{z_1}, one deduces that

(5.27) $\quad P_{z_1}[\mathsf{H}_{V_I} < \mathsf{T}_{\tilde{B}}] = \sum_{x=(y,z) \in V_I} \frac{w_x}{|G|} E_{z_1}^{\mathbb{Z}}\left[\int_0^{\infty} \mathbf{1}_{\{\mathsf{Z}_t = z, t < \mathsf{T}_{\tilde{I}}\}} dt\right]$
$$\times P_x[\tilde{H}_x > T_{\tilde{B}}].$$

Since the expected duration of each visit of Z to any point is equal to 1, we also have

(5.28) $\quad E_{z_1}^{\mathbb{Z}}\left[\int_0^{\infty} \mathbf{1}_{\{\mathsf{Z}_t = z, t < \mathsf{T}_{\tilde{I}}\}} dt\right] = E_{z_1}^{\mathbb{Z}}\left[\sum_{n=0}^{T_{\tilde{I}}} \mathbf{1}_{\{Z_n = z\}}\right]$
$$= P_{z_1}^{\mathbb{Z}}[H_z < T_{\tilde{I}}]/P_z^{\mathbb{Z}}[\tilde{H}_z > T_{\tilde{I}}],$$

where we have applied the strong Markov property at H_z and computed the expectation of the geometrically distributed random variable with success parameter $P_z^{\mathbb{Z}}[\tilde{H}_z > T_{\tilde{I}}]$ in the last step. Standard arguments on one-dimensional simple random walk (see for example [15], Section 3.1, (1.7), p. 179) show with (3.4) that the right-hand side of (5.28) is bounded from below by $h_N(1 - cd_N/h_N)$ and from above by $h_N(1 + cd_N/h_N)$. Substituting what we have found into (5.27) and remembering (5.26), we have proved (5.12). □

134   4. Random walks on cylinders and random interlacements

PROOF OF LEMMA 5.3. In order to prove (5.13) it suffices to show that

(5.29) $\quad \lim_N \max_{m \in \mathcal{J}_I, \mathbf{x} \in \mathbb{V}_m} \left| P_{\Phi_m^{-1}(\mathbf{x})}[T_{\tilde{B}} < \tilde{H}_{V_I}] - \mathbb{P}_{\mathbf{x}}^m[\tilde{H}_{\mathbb{V}_m} = \infty] \right| = 0.$

Indeed, since the sets V_m are disjoint by (5.8) and (5.9), assertion (5.29) implies that

$$\max_{I \in \mathcal{I}} \left| \mathrm{cap}_{\tilde{B}}(V_I) - \sum_{m \in \mathcal{J}_I} \mathrm{cap}^m(\mathbb{V}_m) \right|$$
$$= \max_{I \in \mathcal{I}} \left| \sum_{m \in \mathcal{J}_I} \sum_{\mathbf{x} \in \mathbb{V}_m} \left(P_{\Phi_m^{-1}(\mathbf{x})}[T_{\tilde{B}} < \tilde{H}_{V_I}] - \mathbb{P}_{\mathbf{x}}^m[\tilde{H}_{\mathbb{V}_m} = \infty] \right) w_{\mathbf{x}} \right| \longrightarrow 0,$$

as $N \to \infty$. The statement (5.29) follows from the two claims

(5.30) $\quad \lim_N \max_{m \in \mathcal{J}_I, \mathbf{x} \in \mathbb{V}_m} \left| P_{\Phi_m^{-1}(\mathbf{x})}[T_{\tilde{B}} < \tilde{H}_{V_I}] - P_{\Phi_m^{-1}(\mathbf{x})}[T_{B_m} < \tilde{H}_{V_m}] \right| = 0$

and

(5.31) $\quad \lim_N \max_{m \in \mathcal{J}_I, \mathbf{x} \in \mathbb{V}_m} \left| \mathbb{P}_{\mathbf{x}}^m[\tilde{H}_{\mathbb{V}_m} = \infty] - P_{\Phi_m^{-1}(\mathbf{x})}[T_{B_m} < \tilde{H}_{V_m}] \right| = 0.$

We first prove (5.30). It follows from the inclusions (5.9) that $P_{\Phi_m^{-1}(\mathbf{x})}$-a.s.,

$$T_{\tilde{B}} = T_{B_m} + T_{C_m \times \tilde{I}} \circ \theta^X_{T_{B_m}} + T_{\tilde{B}} \circ \theta^X_{T_{C_m \times \tilde{I}}} \circ \theta^X_{T_{B_m}}.$$

Since the sets B_m are disjoint (cf. (5.8)), the strong Markov property applied at the exit times of B_m and $C_m \times \tilde{I}$ shows that for $x = \Phi_m^{-1}(\mathbf{x}) \in V_m$,

(5.32)
$$P_x[T_{\tilde{B}} < \tilde{H}_{V_I}]$$
$$= E_x \left[T_{B_m} < \tilde{H}_{V_m}, E_{X_{T_{B_m}}} \left[T_{C_m \times \tilde{I}} < H_{V_{I,m}}, P_{X_{T_{C_m \times \tilde{I}}}}[T_{\tilde{B}} < H_{V_I}] \right] \right]$$
$$\geq P_x[T_{B_m} < \tilde{H}_{V_m}] \inf_{x_0 \in \partial B_m} P_{x_0}[T_{C_m \times \tilde{I}} < H_{V_{I,m}}]$$
$$\times \inf_{x_0 \in \partial(C_m \times \tilde{I})} P_{x_0}[T_{\tilde{B}} < H_{V_I}].$$

We now show that a_1 and a_2 tend to 1 as N tends to infinity, where we have set

(5.33) $\quad a_1 = \inf_{x_0 \in \partial B_m} P_{x_0}[T_{C_m \times \tilde{I}} < H_{V_{I,m}}],$
$\quad\quad\quad a_2 = \inf_{x_0 \in \partial(C_m \times \tilde{I})} P_{x_0}[T_{\tilde{B}} < H_{V_I}].$

Concerning a_1, note first that

(5.34) $\quad a_1 \geq 1 - M \max_{m': C_{m'} = C_m} \sup_{x_0 \in \partial B_m} P_{x_0}[H_{V_{m'}} < T_{C_{m'} \times \tilde{I}}].$

With the strong Markov property applied at the entrance time of $\bar{B}_{m'}$, recall that $\bar{B}_m$ is either identical to or disjoint from $\bar{B}_m$ by (5.8), we can replace ∂B_m by $\partial B_{m'}$ on the right-hand side of (5.34). With this remark and the application of the isomorphism $\Psi_{m'}^{z_{m'}}$, one finds with (2.13) and $\hat{o}_m = \psi_m(y_m)$ that

$$\sup_{x_0 \in \partial B_m} P_{x_0}[H_{V_{m'}} < T_{C_{m'} \times \tilde{I}}]$$

$$\leq \sup_{x_0 \in \partial B((\hat{o}_{m'},0),r_N-1)} \hat{\mathbb{P}}_{x_0}^{m'}[H_{\Psi_{m'}^{z_{m'}}(V_{m'})} < T_{\Psi_{m'}^{z_{m'}}(C_{m'} \times \tilde{I})}]$$

$$\leq \sup_{x_0 \in \partial B((\hat{o}_{m'},0),r_N-1)} \hat{\mathbb{P}}_{x_0}^{m'}[H_{\Psi_{m'}^{z_{m'}}(V_{m'})} < \infty].$$

From $\Psi_m^{z_m}(V_m) \subset \Psi_m^{z_m}(B(x_m,\kappa)) = B((\hat{o}_m,0),\kappa)$, see (5.9), and the left-hand estimate in (2.14), we see that the right-hand side tends to 0, and hence a_1 tends to 1 as N tends to infinity. We now show that a_2 tends to 1 as well. The infimum defining a_2 can only be attained for points $x_0 = (y_0, z_0)$ with $y_0 \in \partial C_m$ (if $z_0 \in \partial \tilde{I}$, the probability is equal to 1). Hence, we see that

(5.35) $\quad a_2 \geq 1 - |V_I| \max_{m' \in \mathcal{J}_I} \max_{x' \in \mathbb{V}_{m'}} \sup_{y_0 \in \partial C_m, z_0 \in \tilde{I}} P_{(y_0,z_0)}[H_{\Phi_{m'}^{-1}(x')} < T_{\tilde{B}}].$

By applying the strong Markov property at the entrance time of the set $C_{m'} \times \tilde{I}$ (which is either identical to or disjoint from $C_m \times \tilde{I}$ by (5.10)), it follows that the supremum on the right-hand side of (5.35) is bounded from above by

$$\sup_{y_0 \in \partial (C_{m'}^c), z_0 \in \tilde{I}} P_{(y_0,z_0)}[H_{\Phi_{m'}^{-1}(x')} < T_{\tilde{B}}],$$

which tends to 0 by the estimate (5.25) of Lemma 5.5. Thus, both a_1 and a_2 in (5.33) tend to 1 as N tends to infinity. With (5.32) and the P_x-a.s. inclusion $\{T_{\tilde{B}} < \tilde{H}_{V_I}\} \subseteq \{T_{B_m} < \tilde{H}_{V_m}\}$, we have shown the announced claim (5.30).

To show (5.31), we apply the strong Markov property at the exit time of $\mathbb{B}_m$ and obtain for any $\mathbf{x} \in \mathbb{V}_m \subset \mathbb{B}_m$,

$$\mathbb{P}_\mathbf{x}^m[\tilde{H}_{\mathbb{V}_m} = \infty] = \mathbb{E}_\mathbf{x}^m[T_{\mathbb{B}_m} < \tilde{H}_{\mathbb{V}_m}, \mathbb{P}_{X_{T_{\mathbb{B}_m}}}^m[H_{\mathbb{V}_m} = \infty]].$$

The right-hand side can be bounded from above by

$$\mathbb{P}_\mathbf{x}^m[T_{\mathbb{B}_m} < \tilde{H}_{\mathbb{V}_m}] = P_{\Phi_m^{-1}(\mathbf{x})}[T_{B_m} < \tilde{H}_{V_m}], \text{ cf. (2.12)},$$

and using $\mathbb{V}_m \subset B((o_m,0),\kappa)$ (cf. (5.9)) from below by

$$P_{\Phi_m^{-1}(\mathbf{x})}[T_{B_m} < \tilde{H}_{V_m}](1 - |V_m| \sup_{x_0 \in \partial \mathbb{B}_m} \sup_{\mathbf{x}' \in B((o_m,0),\kappa)} \mathbb{P}_{x_0}^m[H_{\mathbf{x}'} < \infty]).$$

The right-hand estimate in (2.14) shows that this last supremum tends to 0, hence (5.31). This completes the proof of Lemma 5.3. □

PROOF OF LEMMA 5.4. Following the argument of Lemma 4.1 in [36], we begin with the proof of (5.14). To this end, it suffices to show that for

(5.36) $\qquad\qquad \gamma = \mathbf{t}_N \sigma_N^{3/4}$, cf. (3.9), (3.10),

136   4. Random walks on cylinders and random interlacements

and some constant $c_2 > 0$,

(5.37) $$\sup_{z \in \mathbb{Z}} P_z^{\mathbb{Z}}[\mathsf{D}_{k^*-k_*} > c_2\gamma] \xrightarrow{N \to \infty} 0 \text{ and}$$

(5.38) $$\sup_{z \in \mathbb{Z}, x \in G \times \mathbb{Z}} P_z[\mathsf{H}_x \leq c_2\gamma] \xrightarrow{N \to \infty} 0.$$

Observe first that by the definition of the grid in (3.3), the random variables T_O and R_1 are both bounded from above by an exit-time $T_{[z-ch_N, z+ch_N]}$, $P_z^{\mathbb{Z}}$-a.s. With $E_z^{\mathbb{Z}}[T_{[z-ch_N, z+ch_N]}] \leq ch_N^2 \leq c\mathsf{t}_N$, it follows from Khaśminskii's Lemma (see [**33**], Lemma 1.1, p. 292, and also [**18**]) that for some constant $c_3 > 0$,

(5.39) $$\sup_{z \in \mathbb{Z}} E_z^{\mathbb{Z}}\big[\exp\{c_3(T_O \vee R_1)/\mathsf{t}_N\}\big] \leq 2.$$

With the exponential Chebyshev inequality and the strong Markov property applied at the times $R_{k^*-k_*}, D_{k^*-k_*-1}, \ldots, D_1, R_1$, one deduces that

$$\sup_{z \in \mathbb{Z}} P_z^{\mathbb{Z}}[D_{k^*-k_*} > c\gamma] \leq \exp\{-cc_3\sigma_N^{3/4}\} \sup_{z \in \mathbb{Z}} E_z^{\mathbb{Z}}\big[\exp\{c_3 D_{k^*-k_*}/\mathsf{t}_N\}\big]$$

$$\leq \exp\{-cc_3\sigma_N^{3/4}\} \Big(\sup_{z \in \mathbb{Z}} E_z^{\mathbb{Z}}\big[\exp\{c_3(T_O \vee R_1)/\mathsf{t}_N\}\big]\Big)^{2(k^*-k_*)}$$

$$\stackrel{(5.39)}{\leq} \exp\{-cc_3\sigma_N^{3/4} + 2(\log 2)2[\sigma_N^{3/4}]\}.$$

Hence, the claim (5.37) with D replaced by D follows for a suitably chosen constant c. The claim with D for a slightly larger constant c_2 is then a simple consequence of Lemma 3.5, applied with $a_N = k^* - k_*$.

To prove (5.38), note that the expected amount of time spent by the random walk X at a site x during the time interval $[\mathsf{H}_x, \mathsf{H}_x + 1]$ is bounded from below by $(1 \wedge \sigma_1^{\mathsf{X}}) \circ \theta_{\mathsf{H}_x}$. Hence, for $z \in \mathbb{Z}$ and $x = (y', z') \in G \times \mathbb{Z}$, the Markov property at time H_x yields

$$E_z\Big[\int_0^{c_2\gamma+1} \mathbf{1}_{\{\mathsf{X}_t = x\}} dt\Big] \geq P_z[\mathsf{H}_x \leq c_2\gamma] \inf_{x' \in G \times \mathbb{Z}} E_{x'}[1 \wedge \sigma_1^{\mathsf{X}}] \stackrel{A1}{\geq} cP_z[\mathsf{H}_x \leq c_2\gamma].$$

Using the fact that Y_t is distributed according to the uniform distribution on G under P_z, and the bound (2.18) on the heat kernel of Z, the left-hand side is bounded by

$$\frac{c}{|G|} \int_0^{c_2\gamma+1} P_z^{\mathbb{Z}}[\mathsf{Z}_t = z'] dt \leq c\frac{\sqrt{\gamma}}{|G|}.$$

We have therefore found that

$$\sup_{z \in \mathbb{Z}, x \in E} P_z[\mathsf{H}_x \leq c_2\gamma] \leq c\sqrt{\gamma}|G|^{-1}$$

$$\stackrel{(5.36)}{\leq} c\sqrt{\mathsf{t}_N}\sigma^{3/8}|G|^{-1}$$

$$\stackrel{(3.10),(3.9)}{\leq} c(\alpha)(h_N/|G|)^{1/4},$$

and by (3.4) and A2, we know that $h_N/|G|$ is bounded by $|G|^{-\epsilon/4}$. This completes the proof of (5.38) and hence (5.14).

Note that (5.15) is a direct consequence of (5.14), since the probability in (5.15) is smaller than $(\sum_m |V_m|) \sup_{z \in \mathbb{Z}, x \in E} P_z[\mathsf{H}_x \leq \mathsf{D}_1]$.

Finally, the expectation in (5.16) is smaller than

$$P\big[\theta_{\mathsf{D}_{k_*}}^{-1}\{\mathsf{H}_{\cup_I V_I} \leq \mathsf{D}_{k^*-k_*}\}\big] = E\big[P_{\mathsf{Z}_{\mathsf{D}_{k_*}}}[\mathsf{H}_{\cup_I V_I} \leq \mathsf{D}_{k^*-k_*}]\big],$$

and hence (5.16) follows from (5.15). □

6. Estimates on the jump process

In this section, we provide estimates on the jump process $\eta^\mathsf{X} = \eta^\mathsf{Y} + \eta^\mathsf{Z}$ of X that will be of use in the reduction of Theorem 1.1 to the continuous-time result Theorem 5.1 in the next section. There, the number $[\alpha|G|^2]$ of steps of X will be replaced by a random number $\eta^\mathsf{X}_{\alpha'|G|^2}$ of jumps and this will make the local time $L^z(\eta^\mathsf{X}_{\alpha|G|^2})$ appear. We hence prove results on the large N behavior of $\eta^X_{\alpha|G|^2}$ (Lemma 6.4) and $L^z(\eta^\mathsf{X}_{\alpha|G|^2})$ (Lemma 6.5), for $\alpha > 0$. Of course, there is no difficulty in analyzing the Poisson process η^Z of constant parameter 1. The crux of the matter is the N-dependent and inhomogeneous component η^Y. Let us start by investigating the expectation of η^Y_t.

LEMMA 6.1.

(6.1) $$\sup_{y \in G} E^G_y[\eta^\mathsf{Y}_t] \leq \max_{y \in G} w_y t, \text{ and}$$

(6.2) $$E^G[\eta^\mathsf{Y}_t] = tw(G)/|G|, \text{ for } t \geq 0 \text{ and all } N.$$

PROOF. Under P^G_y, $y \in G$, the process

(6.3) $$M_t = \eta_t - \int_0^t w(\mathsf{Y}_s)\,ds,\ t \geq 0,$$

is a martingale, see Chou and Meyer [8], Proposition 3. A proof of a slightly more general fact is also given by Darling and Norris [9], Theorem 8.4. In order to prove (6.1), we take the E^G_y-expectation in (6.3). If we take the E^G-expectation in (6.3) and use that $E^G[w(\mathsf{Y}_s)] = E^G[w(\mathsf{Y}_0)] = w(G)/|G|$ by stationarity, we find (6.2). □

We next bound the covariance and variance of increments of η^Y. Let us denote the compensated increments of η^Y as

(6.4) $$I^\mathsf{Y}_{s,t} = \eta^\mathsf{Y}_t - \eta^\mathsf{Y}_s - (t-s)w(G)/|G|, \text{ for } 0 \leq s \leq t.$$

LEMMA 6.2. Assuming A1, one has for $0 \leq s \leq t \leq s' \leq t'$,

(6.5) $$|\mathrm{cov}_{P^G}(I^\mathsf{Y}_{s,t}, I^\mathsf{Y}_{s',t'})| \leq c_1^2(t-s)(t'-s')|G|\exp\{-(s'-t)\lambda_N\},$$

(6.6) $$\mathrm{var}_{P^G}(I^\mathsf{Y}_{s,t}) \leq c_1(t-s) + c_1^2(t-s)^2.$$

PROOF. In Lemma 6.1, we have proved that $E^G[I_{r,r'}] = 0$ for $0 \le r \le r'$, so that by the Markov property applied at time s', the left-hand side of (6.5) can be expressed as

$$|E^G[I_{s,t}I_{s',t'}]| = |E^G[I_{s,t}(E^G_{Y_{s'}}[I_{0,t'-s'}] - E^G[I_{0,t'-s'}])]|.$$

With an application of the Markov property at time t, this last expression becomes

$$\left|\sum_{y\in G} E^G\big[I_{s,t}(q^G_{s'-t}(Y_t,y) - |G|^{-1})\big]E^G_y[I_{0,t'-s'}]\right|$$

$$\le \sum_{y\in G} E^G\big[|I_{s,t}||q^G_{s'-t}(Y_t,y) - |G|^{-1}|\big]|E^G_y[I_{0,t'-s'}]|.$$

The claim (6.5) thus follows by applying the estimate (4.1) inside the expectation, then (6.1) and $w(G)/|G| \le c_1$ in order to bound the remaining terms.

To show (6.6), we apply the Markov property at time s and domination of η^Y_{t-s} by a Poisson random variable of parameter $c_1(t-s)$ (cf. (2.16)):

$$\mathrm{var}_{P^G}(I^Y_{s,t}) \le E^G[(\eta^Y_t - \eta^Y_s)^2] = E^G[(\eta^Y_{t-s})^2] \le c_1(t-s) + c_1^2(t-s)^2.$$

□

In the next Lemma, we transfer some of the previous estimates to the process $\eta^Y_{\sigma^Z_\cdot}$.

LEMMA 6.3. *Assuming A1*,

(6.7) $\qquad E[\eta^Y_{\sigma^Z_1}] = w(G)/|G|.$

(6.8) $\qquad \sup_{x\in G\times\mathbb{Z}} E_x[\eta^Y_{\sigma^Z_1}] \le c_1.$

(6.9) $\qquad \sup_{x\in G\times\mathbb{Z}} E_x[(\eta^Y_{\sigma^Z_1})^2] \le c_1 + 2c_1^2.$

PROOF. All three claims are shown by using independence of η^Y and σ^Z and applying Fubini's theorem. To show (6.7), note that

$$E[\eta^Y_{\sigma^Z_1}] = E\big[E^G[\eta^Y_t]\big|_{t=\sigma^Z_1}\big] \stackrel{(6.2)}{=} E[\sigma^Z_1]w(G)/|G| = w(G)/|G|.$$

The statements (6.8) and (6.9) are shown similarly, using additionally stochastic domination of η^Y_t by a Poisson random variable of parameter $c_1 t$ (cf. (2.16)). □

We now come to the two main results of this section. As announced, we now analyze the asymptotic behavior of $\eta^X_{\alpha|G|^2}$, where the whole difficulty comes from the component $\eta^Y_{\alpha|G|^2}$. The method we use is to split the time interval $[0,\alpha|G|^2]$ into $[|G|^{\epsilon/2}]$ increments of length longer than λ_N^{-1}. This is possible by A2 and ensures that the bound from (6.5) on the covariance between different increments of η^Y becomes

6. Estimates on the jump process

useful for non-adjacent increments. The following lemma follows from the second moment Chebyshev inequality and the covariance bound applied to pairs of non-adjacent increments.

LEMMA 6.4. *Assuming A1 and (1.7)*,

(6.10) $\quad \lim_N E\big[|\eta^X_{\alpha|G|^2}/(\alpha|G|^2) - (1+\beta)| \wedge 1\big] = 0, \ \text{for } \alpha > 0.$

PROOF. The law of large numbers implies that $\eta^Z_{\alpha|G|^2}/(\alpha|G|^2)$ converges to 1, P_0^Z-a.s. (see, for example [**15**], Chapter 1, Theorem 7.3). Moreover, $\lim_N w(G)/|G| = \beta$ by (1.7). Since $\eta^X = \eta^Y + \eta^Z$, it hence suffices to show that

(6.11) $\quad \lim_N E^G\big[(|\eta^Y_{\alpha|G|^2}/(\alpha|G|^2) - w(G)/|G||) \wedge 1\big] = 0.$

To this end, put $a = [|G|^{\epsilon/2}]$, $\tau = \alpha|G|^2/a$, and write

(6.12) $\quad \eta^Y_{\alpha|G|^2} - \alpha|G|^2(w(G)/|G|) = \sum_{\substack{1 \leq n \leq a \\ n \text{ even}}} I^Y_{(n-1)\tau, n\tau} + \sum_{\substack{1 \leq n \leq a \\ n \text{ odd}}} I^Y_{(n-1)\tau, n\tau}$
$\stackrel{(\text{def.})}{=} \Sigma_1 + \Sigma_2,$

for I^Y as in (6.4). Fix any $\delta > 0$ and $\Sigma \in \{\Sigma_1, \Sigma_2\}$. By Chebyshev's inequality,

(6.13) $\quad P^G[|\Sigma| \geq \delta\alpha|G|^2] \leq \dfrac{1}{\delta^2\alpha^2|G|^4} E^G[\Sigma^2]$
$= \dfrac{1}{\delta^2\alpha^2|G|^4}\Big(\sum_i E^G[(I^Y_{(i-1)\tau, i\tau})^2] + \sum_{i \neq j} E^G[I^Y_{(i-1)\tau, i\tau} I^Y_{(j-1)\tau, j\tau}]\Big),$

where the two sums are over unordered indices i and j in $\{1, \ldots, a\}$ that are either all even or all odd, depending on whether Σ is equal to Σ_1 or to Σ_2. The right-hand side of (6.13) can now be bounded with the help of the estimates on the increments of η^Y in Lemma 6.2. Indeed, with (6.6), the first sum is bounded by $ca\tau^2 \leq c(\alpha)|G|^{4-\epsilon/2}$. For the second sum, we observe that $|i - j| \geq 2$ for all indices i and j, apply (6.5) and A2 and bound the sum with $(|G|\tau)^c \exp\{-c(\alpha)\tau\lambda_N\} \leq |G|^c \exp\{-c(\alpha)|G|^{\epsilon/2}\}$. Hence we find that

$P^G[|\Sigma| \geq \delta\alpha|G|^2] \leq c(\alpha, \delta)(|G|^{-\epsilon/2} + |G|^c \exp\{-c(\alpha)|G|^{\epsilon/2}\}) \to 0,$

as $N \to \infty$, from which we deduce with (6.12) that for our arbitrarily chosen $\delta > 0$,

$P^G\big[|\eta^Y_{\alpha|G|^2}/(\alpha|G|^2) - w/|G|| \geq 2\delta\big] \leq P^G[|\Sigma_1| \geq \delta\alpha|G|^2]$
$+ P^G[|\Sigma_2| \geq \delta\alpha|G|^2] \to 0,$

as N tends to infinity, showing (6.11). This completes the proof of Lemma 6.4. $\square$

140   4. Random walks on cylinders and random interlacements

In the final lemma of this section, we apply a similar analysis to the local time of the process $\pi_\mathbb{Z}(X)$ evaluated at time $\eta^X_{\alpha|G|^2}$. The proof is similar to the preceding argument, although the appearance of η^Y evaluated at the random times σ^Z_n complicates matters. We recall the notation L and $\hat{L}$ for the local times of $\pi_\mathbb{Z}(X)$ and Z from (1.6) and (2.6).

LEMMA 6.5. *Assuming A1, A2 and (1.7),*

(6.14) $\quad \limsup\limits_{N} \sup\limits_{z\in\mathbb{Z}} E\big[(|L^{z,X}_{\eta^X_{\alpha|G|^2}} - (1+\beta)\hat{L}^z_{\eta^Z_{\alpha|G|^2}}|/|G|) \wedge 1\big] = 0, \quad \text{for } \alpha > 0.$

PROOF. Set $T = \alpha|G|^2$. By independence of η^Z and Z, we have

$$E[\hat{L}_{\eta^Z_T}] = E\Big[\sum_{n\geq 0} \mathbf{1}_{\{n<\eta^Z_T\}} P^{\mathbb{Z}}_0[Z_n = z]\Big] \stackrel{(2.20)}{\leq} cE\big[\sqrt{\eta^Z_T}\big] \stackrel{\text{(Jensen)}}{\leq} c(\alpha)|G|.$$

From this estimate and the assumption $w(G)/|G| \to \beta$ made in (1.7), it follows that it suffices to prove (6.14) with $w(G)/|G|$ in place of β. It follows from the definition of L^z in (1.6) that

$$\sum_{n=0}^{\eta^Z_T - 1} \mathbf{1}_{\{Z_n = z\}}(1 + \eta^Y_{\sigma^Z_{n+1}} - \eta^Y_{\sigma^Z_n}) \leq L^{z,X}_{\eta^X_T} \leq \sum_{n=0}^{\eta^Z_T} \mathbf{1}_{\{Z_n = z\}}(1 + \eta^Y_{\sigma^Z_{n+1}} - \eta^Y_{\sigma^Z_n}),$$

hence,

(6.15) $\quad \sup\limits_{z\in\mathbb{Z}} E\Big[\Big|L^{z,X}_{\eta^X_T} - \sum_{n=0}^{\eta^Z_T - 1} \mathbf{1}_{\{Z_n = z\}}(1 + \eta^Y_{\sigma^Z_{n+1}} - \eta^Y_{\sigma^Z_n})\Big|\Big]$
$\leq 1 + E[\eta^Y_{\sigma^Z_{\eta^Z_T+1}} - \eta^Y_{\sigma^Z_{\eta^Z_T}}].$

By independence of η^Y and (σ^Z, η^Z) and the simple Markov property (under P^G) applied at time $\sigma^Z_{\eta^Z_T}$, the expectation on the right-hand side is with (6.1) bounded by $cE[\sigma^Z_{\eta^Z_T+1} - \sigma^Z_{\eta^Z_T}]$. This last expectation is equal to the sum of two independent $\exp(1)$-distributed random variables, so it follows that the right-hand side of (6.15) is bounded by a constant. By these observations, the proof will be complete once we show that

(6.16) $\quad \limsup\limits_{N} \sup\limits_{z\in\mathbb{Z}} E\Big[\Big(\Big|\sum_{n=0}^{\eta^Z_T - 1} \mathbf{1}_{\{Z_n = z\}} S_n\Big|/|G|\Big) \wedge 1\Big] = 0, \quad \text{where}$
$S_n = \eta^Y_{\sigma^Z_{n+1}} - \eta^Y_{\sigma^Z_n} - w(G)/|G|, \quad \text{for } n \geq 0.$

6. Estimates on the jump process

To this end, we will prove that

(6.17) $\quad \limsup_{N} E\Big[\Big(\Big|\sum_{n=0}^{\eta_T^Z-1} \mathbf{1}_{\{Z_n=z\}} S_n - \sum_{n=0}^{[T]} \mathbf{1}_{\{Z_n=z\}} S_n\Big|/|G|\Big) \wedge 1\Big] = 0,$

and

(6.18) $\quad \limsup_{N} E\Big[\Big|\sum_{n=0}^{[T]} \mathbf{1}_{\{Z_n=z\}} S_n\Big|/|G|\Big] = 0.$

In order to show (6.17), we note that by the Chebyshev inequality,

(6.19) $\quad P[|\eta_T^Z - T| \geq T^{3/4}] \leq cT^{-3/2} E[(\eta_T^Z - T)^2] = T^{-1/2}.$

The expectation in (6.17), taken on the complement of the event $\{|\eta_T^Z - T| \geq T^{3/4}\}$, is bounded by

(6.20) $\quad \dfrac{1}{|G|} \sum_{T-cT^{3/4} \leq n \leq T+cT^{3/4}} E[\mathbf{1}_{\{Z_n=z\}}|S_n|].$

Using independence of Z and $\eta_{\sigma z}^Y$ and the heat-kernel bound (2.20), we find that the last expectation is bounded by $cE[|S_n|]/\sqrt{n}$, which by the strong Markov property applied at time σ_n^z, (6.7) and A1 is bounded by $c/\sqrt{n}$. The expression in (6.20) is thus bounded by $cT^{3/8}/|G| = c\alpha|G|^{-1/4}$ and with (6.19), we have proved (6.17).

We now come to (6.18). By the Cauchy-Schwarz inequality, we have for all $z \in \mathbb{Z}$,

(6.21) $\quad E\Big[\Big|\sum_{n=0}^{[T]} \mathbf{1}_{\{Z_n=z\}} S_n\Big|/|G|\Big]^2 \leq \dfrac{1}{|G|^2} E\Big[\Big|\sum_{n=0}^{[T]} \mathbf{1}_{\{Z_n=z\}} S_n\Big|^2\Big].$

We will now expand the square and respectively sum over identical indices, indices of distance at most $[|G|^{2-\epsilon/2}]$, indices of distance greater than $[|G|^{2-\epsilon/2}]$. Proceeding in this fashion, the right-hand side of (6.21) equals

(6.22) $\quad \dfrac{1}{|G|^2}\Big(\sum_{0 \leq n \leq T} E\big[Z_n = z, S_n^2\big]$

$\qquad + 2 \sum_{0 \leq n < n' \leq (n+b) \wedge [T]} E\big[Z_n = Z_{n'} = z, S_n S_{n'}\big]$

$\qquad + 2 \sum_{0 \leq n,\, n+b < n' \leq [T]} E\big[Z_n = Z_{n'} = z, S_n S_{n'}\big]\Big),$

where $b = [|G|^{2-\epsilon/2}]$.

We now treat each of these three sums separately, starting with the first one. By the strong Markov property, (6.9) and A1,

$$(6.23) \qquad \sum_{0 \leq n \leq [T]} E\big[Z_n = z, S_n^2\big] = \sum_{0 \leq n \leq [T]} E\big[Z_n = z, E_{X_{\sigma_n^Z}}[S_0^2]\big]$$
$$\leq c \sum_{0 \leq n \leq [T]} P[Z_n = z].$$

By the heat-kernel bound (2.20), this last sum is bounded by $\sum_n c/\sqrt{n} \leq c\sqrt{T}$. We have thus found that

$$(6.24) \qquad \sum_{0 \leq n \leq [T]} E\big[Z_n = z, S_n^2\big] \leq c(\alpha)|G|.$$

For the second sum in (6.22), we proceed in a similar fashion. The strong Markov property applied at time $\sigma_{n'}^Z \geq \sigma_{n+1}^Z$ and the estimate (6.8) together yield

$$\sum_{0 \leq n < n' \leq (n+b) \wedge [T]} E\big[Z_n = Z_{n'} = z, S_n S_{n'}\big]$$
$$= \sum_{n,n'} E\big[Z_n = Z_{n'} = z, S_n E_{X_{\sigma_{n'}^Z}}[S_0]\big]\Big|$$
$$\leq c \sum_{0 \leq n \leq [T]} E\Big[Z_n = z, |S_n| \sum_{n'=n+1}^{n+b} \mathbf{1}_{\{Z_{n'}=z\}}\Big].$$

Applying the strong Markov property at time σ_{n+1}^Z, we bound the right-hand side by

$$c \sum_{0 \leq n \leq [T]} \Big(E[Z_n = z, |S_n|] \sum_{n'=0}^{b-1} \sup_{z' \in \mathbb{Z}} P_{z'}^{\mathbb{Z}}[Z_{n'} = z] \Big)$$
$$\stackrel{(2.20)}{\leq} c\sqrt{b} \sum_{0 \leq n \leq [T]} E[Z_n = z, |S_n|].$$

The sum on the right-hand side can be bounded by $c(\alpha)|G|$ with the same arguments as in (6.23)-(6.24), the only difference being the use of the estimate (6.8) rather than (6.9). Inserting the definition of b from (6.22), we then obtain

$$(6.25) \qquad \sum_{0 \leq n < n' \leq n+b} E\big[Z_n = Z_{n'} = z, S_n S_{n'}\big] \leq c(\alpha)|G|^{2-\epsilon/4}.$$

7. Proof of the result in discrete time

For the expectation in the third sum in (6.22), we first use independence of Z and S, then (6.7) and the fact that the process σ^Z has iid exp(1)-distributed increments for the second line and thus obtain

$$\left|E\left[Z_n = Z_{n'} = z, S_n S_{n'}\right]\right| = P[Z_n = Z_{n'} = z]\left|E[S_n S_{n'}]\right| \leq \left|E[S_n S_{n'}]\right|$$
$$= \left|E\left[(\eta^Y_{\sigma^Z_{n+1}} - \eta^Y_{\sigma^Z_n})(\eta^Y_{\sigma^Z_{n'+1}} - \eta^Y_{\sigma^Z_{n'}})\right]\right.$$
$$\left. - \frac{w(G)^2}{|G|^2} E\left[(\sigma^Z_{n+1} - \sigma^Z_n)(\sigma^Z_{n'+1} - \sigma^Z_{n'})\right]\right|.$$

Independence of η^Y and σ^Z and an application of Fubini's theorem then allows to bound the the third sum in (6.22) by

$$\sum_{0 \leq n, \, n+b < n' \leq [T]} \left|E^{\mathbb{Z}}_0[h(\sigma^Z_n, \sigma^Z_{n+1}, \sigma^Z_{n'}, \sigma^Z_{n'+1})]\right|,$$

where $h(s, t, s', t') = \mathrm{cov}_{P^G}(\eta^Y_t - \eta^Y_s, \eta^Y_{t'} - \eta^Y_{s'})$.

Via the estimate (6.5) on the covariance, this expression is bounded by

$$c|G| \sum_{0 \leq n, \, n+b < n' \leq [T]} E^{\mathbb{Z}}_0\left[(\sigma^Z_{n+1} - \sigma^Z_n)(\sigma^Z_{n'+1} - \sigma^Z_{n'}) \exp\{-(\sigma^Z_{n'} - \sigma^Z_{n+1})\lambda_N\}\right].$$

Since the process σ^Z has iid exp(1)-distributed increments, this sum can be simplified to

$$\sum_{0 \leq n, \, n+b < n' \leq [T]} E\left[\exp\{-\sigma^Z_1 \lambda_N\}\right]^{n'-n-1} \leq \sum_{0 \leq n \leq [T]} \sum_{n' > n+b} \left(\frac{1}{1+\lambda_N}\right)^{n'-n-1}$$
$$= [T]\frac{1+\lambda_N}{\lambda_N}\left(\frac{1}{1+\lambda_N}\right)^b \leq c(\alpha)|G|^c e^{-cb\lambda_N} \overset{A2}{\leq} c(\alpha)|G|^c \exp\{-c|G|^{\epsilon/2}\}.$$

Combining this bound on the third sum in (6.22) with the bounds (6.24) and (6.25) on the first and second sums, we have shown (6.18), hence (6.16). This completes the proof of Lemma 6.5. □

7. Proof of the result in discrete time

In this section, we prove Theorem 1.1. We assume that A1-A10 and (1.7) hold. The proof uses the estimates of the previous section to deduce Theorem 1.1 from the continuous-time version stated in Theorem 5.1.

PROOF OF THEOREM 1.1. The transience of the graphs $\mathbb{G}_m \times \mathbb{Z}$ follows from Theorem 5.1. Consider again finite subsets $\mathbb{V}_m$ of $\mathbb{G}_m \times \mathbb{Z}$, $1 \leq m \leq M$ and set $V_m = \Phi^{-1}_m(\mathbb{V}_m)$. We show that for $\theta_m \in \mathbb{R}_+$, $\alpha > 0$,

(7.1) $$\lim_N E\left[\prod_{1 \leq m \leq M} \mathbf{1}_{\{H_{V_m} > T\}} \exp\left\{-\frac{\theta_m}{|G|} L^{z_m}_T\right\}\right] = B(\alpha),$$

where $T = \alpha|G|^2$ and

$$B(\alpha) = E^W\left[\exp\left\{-\sum_{1 \le m \le M} L(v_m, \alpha/(1+\beta))(\mathrm{cap}^m(\mathbb{V}_m) + (1+\beta)\theta_m)\right\}\right].$$

This implies Theorem 1.1, by the standard arguments described below (5.6). Recall that two sequences are said to be limit equivalent if their difference tends to 0 as N tends to infinity. If we apply Theorem 5.1 with $\alpha/(1+\beta)$ in place of α, we obtain

$$\lim_N E\left[\prod_{1 \le m \le M} \mathbf{1}_{\{H_{V_m} > \eta^{\mathsf{X}}_{T/(1+\beta)}\}} \exp\left\{-\frac{\theta_m(1+\beta)}{|G|} \mathsf{L}^{z_m}_{T/(1+\beta)}\right\}\right] = B(\alpha).$$

By (3.17), the expression on the left-hand side is limit equivalent to the same expression with L replaced by $\hat{L}$. Hence, we have

$$\lim_N E\left[\prod_{1 \le m \le M} \mathbf{1}_{\{H_{V_m} > \eta^{\mathsf{X}}_{T/(1+\beta)}\}} \exp\left\{-\frac{\theta_m(1+\beta)}{|G|} \hat{L}^{z_m}_{T/(1+\beta)}\right\}\right] = B(\alpha).$$

By the law of large numbers, $\lim_N \eta^{\mathsf{Z}}_{T/(1+\beta)}(T/(1+\beta))^{-1} = 1$, P-a.s. Making use of the monotonicity of the left-hand side in the local time and continuity of $B(\cdot)$, we deduce that

$$\lim_N E\left[\prod_{1 \le m \le M} \mathbf{1}_{\{H_{V_m} > \eta^{\mathsf{X}}_{T/(1+\beta)}\}} \exp\left\{-\frac{\theta_m(1+\beta)}{|G|} \hat{L}^{z_m}_{\eta^{\mathsf{Z}}_{T/(1+\beta)}}\right\}\right] = B(\alpha).$$

The estimate (6.14) then shows that the expression on the left-hand side is limit equivalent to the same expression with $(1+\beta)\hat{L}^{z_m}_{\eta^{\mathsf{Z}}_{T/(1+\beta)}}$ replaced by $L^{z_m}_{\eta^{\mathsf{X}}_{T/(1+\beta)}}$, i.e.

$$\lim_N E\left[\prod_{1 \le m \le M} \mathbf{1}_{\{H_{V_m} > \eta^{\mathsf{X}}_{T/(1+\beta)}\}} \exp\left\{-\frac{\theta_m}{|G|} L^{z_m}_{\eta^{\mathsf{X}}_{T/(1+\beta)}}\right\}\right] = B(\alpha).$$

Applying the estimate (6.10), with the same monotonicity and continuity arguments as in the beginning of the proof, we can replace $\eta^{\mathsf{X}}_{T/(1+\beta)}$ by T, hence infer that (7.1) holds. $\square$

8. Examples

In this section, we apply Theorem 1.1 to three examples of graphs G: The d-dimensional box of side-length N, the Sierpinski graph of depth N, and the d-ary tree of depth N ($d \ge 2$). In each case, we check assumptions A1-A10, stated after (2.9). In all examples it is implicitly understood that all edges of the graphs have weight $1/2$. We begin with a lemma from [28] asserting that the continuous-time spectral gap has the same order of magnitude as its discrete-time analog λ^d_N. This result will be useful for checking A2.

LEMMA 8.1. *Assume A1 and let λ^d_N bet the smallest non-zero eigenvalue of the matrix $I - P(G)$, where $P(G) = (p^G(y,y'))$ is the transition*

8. Examples

matrix of Y under P^G. Then there are constants $c(c_0, c_1), c'(c_0, c_1) > 0$ (cf. A1), such that for all N,

$$(8.1) \qquad c(c_0, c_1)\lambda_N^d \leq \lambda_N \leq c'(c_0, c_1)\lambda_N^d.$$

PROOF. We follow arguments contained in [28]. With the Dirichlet form $\mathcal{D}_\pi(.,.)$ defined as $\mathcal{D}_\pi(f, f) = \mathcal{D}_N(f, f)\frac{|G|}{w(G)}$, for $f : G \to \mathbb{R}$ (cf. (2.8)), one has (cf. [28], Definition 2.1.3, p. 327)

$$(8.2) \qquad \lambda_N^d = \min\left\{\frac{\mathcal{D}_\pi(f, f)}{\mathrm{var}_\pi(f)} : f \text{ is not constant}\right\}.$$

From A1, it follows that, for any $f : G \to \mathbb{R}$,

$$(8.3) \qquad \begin{array}{rcccl} c_1^{-1}\mathcal{D}_N(f, f) & \leq & \mathcal{D}_\pi(f, f) & \leq & c_0^{-1}\mathcal{D}_N(f, f), \text{ and} \\ c_0 c_1^{-1}\mu(y) & \leq & \pi(y) & \leq & c_1 c_0^{-1}\mu(y), \quad \text{for any } y \in G. \end{array}$$

Using $\mathrm{var}_\pi(f) = \inf_{\theta \in \mathbb{R}} \sum_{y \in G}(f(y) - \theta)^2 \pi(y)$ and the analogous statement for var_μ, the estimate in the second line implies that

$$(8.4) \qquad c_0 c_1^{-1}\mathrm{var}_\mu(f) \leq \mathrm{var}_\pi(f) \leq c_1 c_0^{-1}\mathrm{var}_\mu(f), \text{ for any } f : G \to \mathbb{R}.$$

Lemma 8.1 then follows by using (8.3) and (8.4) to compare the definition (2.9) of λ_N with the characterization (8.2) of λ_N^d. □

The following lemma provides a sufficient criterion for assumption A10.

LEMMA 8.2. *Assuming A1-A9 and that*

$$(8.5) \qquad \lim_N \sum_{n=1}^{[\lambda_N^{-1}|G|^\epsilon]} \sup_{\substack{y_0 \in \partial(C_m^c) \\ y \in B(y_m, \rho_0)}} p_n^G(y_0, y)\frac{1}{\sqrt{n}} = 0, \text{ for any } \rho_0 > 0,$$

A10 holds as well.

PROOF. For $\mathbf{x} = (\mathbf{y}, z)$, the probability in A10 is bounded from above by

$$(8.6) \qquad \sum_{n=1}^{[\lambda_N^{-1}|G|^\epsilon]} P_{(y_0, z_0)}[Y_n = \phi_m^{-1}(\mathbf{y}), z_m + z \in Z_{[\sigma_n^Y, \sigma_{n+1}^Y]}],$$

using that $y_0 \neq \phi_m^{-1}(\mathbf{y})$ for large N (cf. A6) in order to drop the term $n = 0$. With the same estimates as in the proof of Lemma 2.3, see (2.22)-(2.23), the expression in (8.6) can be bounded by a constant times the sum on the left-hand side of (8.5). □

8.1. The d-dimensional box. The d-dimensional box is defined as the graph with vertices
$$G_N = \mathbb{Z}^d \cap [0, N-1]^d, \text{ for } d \geq 2,$$
and edges between any two vertices at Euclidean distance 1. In contrast to the similar integer torus considered in [**36**], the box admits different limit models for the local pictures, depending on how many coordinates y_m^i of the points y_m are near the boundary.

THEOREM 8.3. *Consider $x_{m,N}$, $1 \leq m \leq M$, in $G_N \times \mathbb{Z}$ satisfying A3 and A4, and assume that for any $1 \leq m \leq M$, there is a number $0 \leq d(m) \leq d$, such that*

(8.7) $\quad y_{m,N}^i \wedge (N - y_{m,N}^i)$ *is constant for $1 \leq i \leq d(m)$ and all large N,*

(8.8) $\quad \lim_N y_{m,N}^i \wedge (N - y_{m,N}^i) = \infty$ *for $d(m) < i \leq d$.*

Then the conclusion of Theorem 1.1 holds with $\mathbb{G}_m = \mathbb{Z}_+^{d(m)} \times \mathbb{Z}^{d-d(m)}$ and $\beta = d$.

PROOF. We check that assumptions A1-A10 and (1.7) are satisfied and apply Theorem 1.1. Assumption A1 is checked immediately. With Lemma 8.1 and standard estimates on λ_N^d for simple random walk on $[0, N-1]^d$ (cf. [**28**], Example 2.1.1. on p. 329 and Lemma 2.2.11, p. 338), we see that $cN^{-2} \leq \lambda_N$, and A2 follows. We have assumed A3 and A4 in the statement. For A5, we define the sequence r_N, the vertices $o_m \in \mathbb{G}_m$ and the isomorphisms ϕ_m by

$$r_N = \frac{1}{4^M 10} \Big(\min_{m \neq m'} |x_m - x_{m'}|_\infty \wedge \min_m \min_{d(m) < i \leq d} (y_m^i \wedge (N - y_m^i)) \wedge N \Big),$$
$$o_m = (y_m^1 \wedge (N - y_m^1), \ldots, y_m^{d(m)} \wedge (N - y_m^{d(m)}), 0, \ldots, 0),$$
$$\phi_m(y) = (y^1 \wedge (N - y^1), \ldots, y^{d(m)} \wedge (N - y^{d(m)}),$$
$$y^{d(m)+1} - y_m^{d(m)+1}, \ldots, y^d - y_m^d).$$

Then $r_N \to \infty$ by A3 and (8.8), o_m remains fixed by (8.7), ϕ_m is an isomorphism from $B(y_m, r_N)$ to $B(o_m, r_N)$ for large N, and A5 follows. Recall that a crucial step in the proof of Theorem 1.1 was to prove that the random walk, when started at the boundary of one of the balls B_m, does not return to the close vicinity of the point x_m before exiting $G \times [-h_N, h_N]$, see Lemma 5.3, (5.33) and below. In the present context, h_N is roughly of order N, see (3.4). However, the radius r_N of the ball B_m can be required to be much smaller if the distances between different points diverge only slowly, cf. (5.1). We therefore needed to assume that larger neighborhoods $C_m \times \mathbb{Z}$ of the points x_m are sufficiently transient by requiring that the sets $\bar{C}_m$ are isomorphic to subsets of suitable infinite graphs $\hat{\mathbb{G}}_m$. In the present context, we

8. Examples

choose $\hat{\mathbb{G}}_m = \mathbb{Z}_+^d$ for all m, see Remark 8.4 below on why a choice different from $\mathbb{G}_m$ is required. We choose the sets C_m with the help of Lemma 3.2. Applied to the points $y_1, \ldots, y_m$, with $a = \frac{1}{4^M 10} N$ and $b = 2$, Lemma 3.2 yields points $y_1^*, \ldots, y_M^*$ (some of them may be identical) and a p between $\frac{1}{4^M 10} N$ and $\frac{1}{10} N$, such that for $1 \leq m \leq M$,

(8.9) either $C_m = C_{m'}$ or $C_m \cap C_{m'} = \emptyset$ for $C_m = B(y_m^*, 2p)$,

and such that the balls with the same centers and radius p still cover $\{y_1, \ldots, y_M\}$. Since $r_N \leq p$, we can associate to any m one of the sets C_m such that A6 is satisfied. The diameter of $\bar{C}_m$ is at most $2N/5 + 3$, so each of the one-dimensional projections $\pi_k(\bar{C}_m)$, $1 \leq k \leq d$, of $\bar{C}_m$ on the d different axes contains at most one of the two integers 0 and $N-1$ for large N. Hence, there is an isomorphism ψ_m from $\bar{C}_m$ into $\mathbb{Z}_+^d$ such that A7 is satisfied. Assumption A8 directly follows from from (8.9). We now turn to A9. By embedding $\mathbb{Z}_+^d$ into $\mathbb{Z}^d$, one has for any $\mathbf{y}$ and $\mathbf{y}'$ in $\mathbb{Z}_+^d$,

$$p_n^{\mathbb{Z}_+^d}(\mathbf{y}, \mathbf{y}') \leq 2^d \sup_{\mathbf{y}, \mathbf{y}' \in \mathbb{Z}^d} p_n^{\mathbb{Z}^d}(\mathbf{y}, \mathbf{y}') \leq c(d) n^{-d/2},$$

using the standard heat kernel estimate for simple random walk on $\mathbb{Z}^d$, see for example [**19**], p. 14, (1.10). Since $d \geq 2$, this is more than enough for A9. In order to check A10, it is sufficient to prove the hypothesis (8.5) of Lemma 8.2. To this end, we compare the probability $P_{y_0}^G$ with $P_{y_0}^{\mathbb{Z}^d}$ under which the canonical process $(Y_n)_{n \geq 0}$ is a simple random walk on $\mathbb{Z}^d$. We define the map $\pi : \mathbb{Z}^d \to G_N$ by $\pi((y_i)_{1 \leq i \leq d}) = (\min_{k \in \mathbb{Z}} |y_i - 2kN|)_{1 \leq i \leq d}$, i.e. in each coordinate, π is a sawtooth map. Then $(Y_n)_{n \geq 0}$ under $P_{y_0}^G$ has the same distribution as $(\pi(Y_n))_{n \geq 0}$ under $P_{y_0}^{\mathbb{Z}^d}$. It follows that for $y_0 \in \partial(C_m^c)$, $y \in B(y_m, \rho_0)$,

(8.10) $$p_n^G(y_0, y) = \sum_{y' \in \mathcal{S}_y} p_n^{\mathbb{Z}^d}(y_0, y'),$$

where $\mathcal{S}_y = 2N\mathbb{Z}^d + \Big\{ \sum_{1 \leq i \leq d} l_i e_i y^i : l \in \{-1, 1\}^d \Big\}$.

The probability in this sum is bounded by

$$\frac{c}{n^{d/2}} \exp\Big\{ \frac{c' |y_0 - y'|^2}{n} \Big\},$$

as follows, for example, from Telcs [**42**], Theorem 8.2 on p. 99, combined with the on-diagonal estimate from the local central limit theorem (cf. [**19**], p. 14, (1.10)). If we insert this bound into (8.10) and split the sum into all possible distances between y_0 and y' (necessarily

148   4. Random walks on cylinders and random interlacements

this distance is at least $p - \rho_0 \geq cN$, cf. (8.9)), we obtain

$$p_n^G(y_0, y) \leq \sum_{k \geq 1} \frac{c}{n^{d/2}} \exp\left\{-\frac{c'k^2N^2}{n}\right\}k^{d-1}$$

$$\leq \frac{c}{n^{d/2}} \int_0^\infty x^{d-1} \exp\left\{-\frac{c'x^2N^2}{n}\right\}dx \leq \frac{c}{N^d}.$$

By $cN^{-2} \leq \lambda_N$, checked under A2 above, this is more than enough to imply (8.5), hence A10. Finally, one immediately checks that (1.7) holds with $\beta = d$. Hence, Theorem 1.1 applies and yields the result. □

REMARK 8.4. In the last proof, we have used the possibility of choosing the auxiliary graphs $\hat{\mathbb{G}}_m$ in assumption A7 different from the graphs $\mathbb{G}_m$ in A5. This is necessary for the following reason: To check assumption A10, we need the diameter of each set $\bar{C}_m$ to be of order N in the above argument. Hence, the set $\bar{C}_m$ can look quite different from the ball $B(y_m, r_N)$. Indeed, $\bar{C}_m$ may touch the boundary of the box G in more dimensions than its much smaller subset $B(y_m, r_N)$. As a result, $\bar{C}_m$ may not to be isomorphic to a neighborhood in the same graph $\mathbb{G}_m$ as $B(y_m, r_N)$. However, our chosen $\bar{C}_m$ is always isomorphic to a neighborhood in $\mathbb{Z}_+^d$ for all m.

8.2. The Sierpinski graph. For $y \in \mathbb{R}^2$ and $\theta \in [0, 2\pi)$, we denote by $\rho_{y,\theta}$ the anticlockwise rotation around y by the angle θ. The vertex-set of the Sierpinski graph G_N of depth N is defined by the following increasing sequence (see also the top of Figure 1):

$$G_0 = \{s_0 = (0,0), s_1 = (1,0), s_2 = \rho_{(0,0),\pi/3}(s_1)\} \subset \mathbb{R}^2,$$
$$G_{N+1} = G_N \cup (\rho_{2^N s_1, 4\pi/3} G_N) \cup (\rho_{2^N s_2, 2\pi/3} G_N), \text{ for } N \geq 0.$$

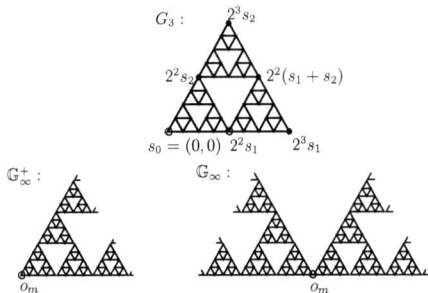

Figure 1: An illustration of G_3 (top) and the infinite limit models $\mathbb{G}_\infty^+$ (bottom left) and $\mathbb{G}_\infty$ (bottom right).

8. Examples

The edge-set of G_N contains an edge between every pair of vertices in G_N at Euclidean distance 1. Note that the vertices in $2^N G_0 \subset G_N$ have degree 2 and all other vertices of G_N have degree 4.

Denoting the reflection around the y-axis by σ, i.e. $\sigma((y_1, y_2)) = (-y_1, y_2)$ for $(y_1, y_2) \in \mathbb{R}^2$, the two-sided infinite Sierpinski graph has vertices

$$\mathbb{G}_\infty = \mathbb{G}_\infty^+ \cup \sigma \mathbb{G}_\infty^+, \text{ where } \mathbb{G}_\infty^+ = \cup_{N \geq 0} G_N,$$

and an edge between any pair of vertices in $\mathbb{G}_\infty^+$ or in $\sigma \mathbb{G}_\infty^+$ at Euclidean distance 1. We refer to the bottom of Figure 1 for illustrations. For $N \geq 0$, we define the surjection $s_N : G_{N+1} \to G_N$ by

$$s_N(y) = \begin{cases} y & \text{for } y \in G_N, \\ \rho_{2^N s_1, 2\pi/3}(y) & \text{for } y \in \rho_{2^N s_1, 4\pi/3}(G_N) \setminus G_N \\ \rho_{2^N s_2, 4\pi/3}(y) & \text{for } y \in \rho_{2^N s_2, 2\pi/3}(G_N) \setminus G_N. \end{cases}$$

We then define the mapping π_N from $\mathbb{G}_\infty^+$ onto G_N by

$$\pi_N(y) = s_N \circ s_{N+1} \circ \cdots \circ s_{m-1}(y) \text{ for } y \in G_m \text{ with } m > N.$$

Note that π_N is well-defined: Indeed, the vertex-sets G_N are increasing in N and if $y \in G_{m_1} \subset G_{m_2}$ for $N < m_1 < m_2$, then $s_k(y) = y$ for $k \geq m_1$, so that $s_N \circ \cdots \circ s_{m_2-1}(y) = s_N \circ \cdots \circ s_{m_1-1}(y)$. We will use the following lemma:

LEMMA 8.5. *For any* $\mathbf{y} \in \mathbb{G}_\infty^+$, *the distribution of the random walk* $(Y_n)_{n \geq 0}$ *under* $\mathbb{P}_{\pi_N(\mathbf{y})}^{G_N}$ *is equal to the distribution of the random walk*

$$(\pi_N(Y_n))_{n \geq 0}$$

under $\mathbb{P}_\mathbf{y}^{\mathbb{G}_\infty^+}$.

PROOF. The result follows from the Markov property once we check that for any $y, y' \in G_N$, $\mathbf{y} \in \mathbb{G}_\infty^+$ with $y = \pi_N(\mathbf{y})$,

(8.11) $$p^{G_N}(y, y') = \sum_{y_1' \in \pi_N^{-1}(y')} p^{\mathbb{G}_\infty^+}(\mathbf{y}, y_1').$$

We choose $m \geq N$ such that $\mathbf{y} \in G_m$. Then the right-hand side equals

$$\sum_{y_1' \in \pi_N^{-1}(y')} p^{G_{m+1}}(\mathbf{y}, y_1') = \sum_{y_1' \in s_N^{-1}(y')} \sum_{y_2' \in s_{N+1}^{-1}(y_1')} \cdots \sum_{y_m' \in s_m^{-1}(y_{m-N}')} p^{G_{m+1}}(\mathbf{y}, y_m').$$

By induction on m, it hence suffices to show that for $y, y' \in G_m$ and $\hat{y} \in s_m^{-1}(y)$,

(8.12) $$p^{G_m}(y, y') = \sum_{y_1' \in s_m^{-1}(y') \cap B(\hat{y}, 1) \subset G_{m+1}} p^{G_{m+1}}(\hat{y}, y_1').$$

If $\hat{y} \in G_{m+1} \setminus \{2^m s_1, 2^m s_2, 2^m(s_1 + s_2)\}$, then (8.12) follows from the observation that s_m maps the distinct neighbors of $\hat{y}$ in G_{m+1} to the distinct neighbors of y in G_m. If $\hat{y} \in \{2^m s_1, 2^m s_2, 2^m(s_1 + s_2)\}$, then $\hat{y}$ has four neighbors in G_{m+1}, two of which are mapped to each of the

150   4. Random walks on cylinders and random interlacements

two neighbors of $y \in \{2^m s_1, 2^m s_2, (0,0)\}$ in G_m and this implies again (8.12). □

In the following theorem, we consider points y_m that are either the corner $(0,0)$ or the vertex $(2^{N-1}, 0)$ and obtain the two different limit models $\mathbb{G}_\infty^+ \times \mathbb{Z}$ and $\mathbb{G}_\infty \times \mathbb{Z}$ for the corresponding local pictures.

THEOREM 8.6. *Consider $0 \leq M' \leq M$ and vertices $x_{m,N}$, $1 \leq m \leq M$, in $G_N \times \mathbb{Z}$ satisfying A3 and A4 and assume that*

(8.13) $y_{m,N} = (0,0)$, *for* $1 \leq m \leq M'$, *and*

$y_{m,N} = (2^{N-1}, 0)$, *for* $M' < m \leq M$.

Then the conclusion of Theorem 1.1 holds with $\mathbb{G}_m = \mathbb{G}_\infty^+$ for $1 \leq m \leq M'$, $\mathbb{G}_m = \mathbb{G}_\infty$ for $M' < m \leq M$ and $\beta = 2$.

PROOF. Let us again check that the hypotheses A1-A10 and (1.7) are satisfied. One easily checks that A1 holds with $c_0 = 1$ and $c_1 = 2$. Using Lemma 8.1 and the explicit calculation of λ_N^d by Shima [**29**], we find that $c5^{-N} \leq \lambda_N \leq c'5^{-N}$. Indeed, in the notation of [**29**], Proposition 3.3 in [**29**] shows that λ_N^d is given by $\phi_-^{(N)}(3)$ for the function ϕ_- defined above Remark 2.16, using our N in place of m and setting the N of [**29**] equal to 3. Then $\lambda_N^d = \phi_-^{(N)}(3)$ is decreasing in N and converges to the fixed point 0 of ϕ_-. With Taylor's theorem it then follows that $\lambda_N^d 5^N$ converges to 1. Since $|G_N| = 3 + \sum_{n=1}^N 3^n \leq c3^N$, A2 holds. We have assumed A3 and A4 in the statement. For A5, we define the radius

$$r_N = \frac{1}{4}(2^{N-1} \wedge \min_{1 \leq m < m' \leq M} d(x_m, x_{m'})), \text{ and set}$$

$o_m = (0,0)$, for all m.

The balls $B(y_m, r_N) \subset G_N$ intersect $2^N G_0$ only at the points y_m, because the distance between different points of $2^N G_0$ equals 2^N. We can therefore define the isomorphisms ϕ_m from $B(y_m, r_N)$ to $B((0,0), r_N) \subset \mathbb{G}_m$ as the identity for $m \leq M'$ and as the translation by $(-2^{N-1}, 0)$ for $m > M'$ and A5 follows. As in the previous example, the radius r_N defined in (5.1) can be small compared with the square root of the relaxation time, so it is essential for the proof that larger neighborhoods $C_m \times \mathbb{Z}$ of the points x_m are sufficiently transient. In the present case, we define the auxiliary graphs as $\hat{\mathbb{G}}_m = \mathbb{G}_m$ and $C_m = B(y_m, 2^{N-1}/3)$ for $1 \leq m \leq M$. Then A6 holds, because $r_N < 2^{N-1}/3$ for large N and the isomorphisms ψ_m required for A7 can be defined in a similar fashion as the isomorphisms ϕ_m above. Assumption A8 is immediate. We now check A9. It is known from [**5**] (see also [**17**]) that for any $\mathbf{y}$ and $\mathbf{y}'$ in $\mathbb{G}_\infty$,

(8.14) $p_n^{\mathbb{G}_\infty}(\mathbf{y}, \mathbf{y}') \leq cn^{-d_s/2} \exp\left\{-c'\left(\frac{d(\mathbf{y}, \mathbf{y}')^{d_w}}{n}\right)^{1/(d_w-1)}\right\},$

8. Examples

for $d_s = 2\log 3/\log 5$, $d_w = \log 5/\log 2$ and $n \geq 1$. Since

(8.15) $\qquad p_n^{\mathbb{G}_\infty^+}(y_0, y) = p_n^{\mathbb{G}_\infty}(y_0, y) + p_n^{\mathbb{G}_\infty}(y_0, \sigma y)$

and $\log 3/\log 5 > 1/2$, this is enough for A9. To prove A10, we use Lemma 8.2 and only check (8.5). To this end, note that $B(y_m, \rho_0) \subseteq \mathcal{K} \subseteq G_N$, for $\mathcal{K} = \cup_{y' \in 2^{N-1}G_1} B(y', \rho_0)$ and that the preimage of the vertices in $2^{N-k}G_k \subset G_N$ under π_N is $2^{N-k}\mathbb{G}_\infty^+$ for $0 \leq k \leq N$. It follows from Lemma 8.5 that for $y_0 \in \partial(C_m^c)$, $y \in B(y_m, \rho_0) \subseteq \mathcal{K}$ and $N \geq c(\rho_0)$,

(8.16) $\qquad p_n^{G_N}(y_0, y) \leq \sum_{y' \in \mathcal{K}} p_n^{G_N}(y_0, y')$

$$= \sum_{y' \in \mathcal{K}} p_n^{\mathbb{G}_\infty^+}(y_0, \mathbf{y}), \text{ for } \mathcal{K} = \bigcup_{y \in 2^{N-1}\mathbb{G}_\infty^+} B(\mathbf{y}, \rho_0).$$

Observe now that for any given vertex $\mathbf{y}'$ in $\mathbb{G}_\infty$, the number of vertices in $B(\mathbf{y}', 2^k) \cap \mathcal{K}$ is less than $c(\rho_0)|B(\mathbf{y}', 2^k) \cap 2^{N-1}\mathbb{G}_\infty^+| \leq c(\rho_0)3^{k-N}$. Also, it follows from the choice of C_m that $d(y_0, 2^{N-1}\mathbb{G}_\infty^+) \geq c2^N$, so the distance between y_0 and any point in $\mathcal{K}$ is at least $c(\rho_0)2^N$. Summing over all possible distances in (8.16), we deduce with the help of (8.14) and (8.15) that

$$p_n^{G_N}(y_0, y) \leq c(\rho_0) \sum_{l=1}^{\infty} 3^l n^{-d_s/2} \exp\left\{-c'(\rho_0)\left(\frac{2^{(N+l)d_w}}{n}\right)^{1/(d_w-1)}\right\}$$

$$\leq c(\rho_0) n^{-d_s/2} \int_0^{\infty} 3^x \exp\left\{-c'(\rho_0)\left(\frac{5^{N+x}}{n}\right)^{1/(d_w-1)}\right\} dx.$$

After substituting $x = y - N + \log n/\log 5$, this expression is seen to be bounded by

$$c(\rho_0)3^{-N} \int_{-\infty}^{\infty} 3^y \exp\left\{-c'(\rho_0)5^{y/(d_w-1)}\right\} dy \leq c(\rho_0)3^{-N}.$$

By $\sqrt{5} < 3$ and $c5^{-N} \leq \lambda_N$, as we have seen under A2, this is more than enough for (8.2), hence A10. Finally, it is straightforward to check that (1.7) holds with $\beta = 2$. Hence, Theorem 1.1 applies and yields the result. □

8.3. The d-ary tree. For a fixed integer $d \geq 2$, we let $\mathbb{G}_o$ be the infinite $d+1$-regular graph without cycles, called the infinite d-ary tree. We fix an arbitrary vertex $o \in \mathbb{G}_o$ and call it the root of the tree. See Figure 2 (left) for a schematic illustration in the case $d = 2$.

We choose G_N as the ball of radius N centered at $o \in \mathbb{G}_o$. For any vertex y in G_N, we refer to the number $|y| = N - d(y, o)$ as the height of y. Vertices in G_N of depth N (or height 0) are called leaves. The boundary-tree $\mathbb{G}_\Diamond$ contains the vertices

$$\mathbb{G}_\Diamond = \{(k; s) : k \geq 0, s \in S_d\},$$

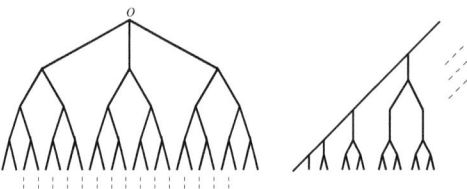

Figure 2: A schematic illustration of $\mathbb{G}_o$ (left) and $\mathbb{G}_\diamond$ (right) for $d = 2$.

where S_d is the set of infinite sequences $s = (s_1, s_2, \ldots)$ in $\{1, \ldots, d\}^{[1,\infty)}$ with at most finitely many terms different from 1. The graph $\mathbb{G}_\diamond$ has edges $\{(k; s), (k+1; s')\}$ for vertices $(k; s)$ and $(k+1; s')$ whenever $s_{n+1} = s'_n$ for all $n \geq 1$. In this case, we refer to the number $k = |(k; s)|$ as the height of the vertex $(k; s)$ and to all vertices at height 0 as leaves. See Figure 2 (right) for an illustration of $\mathbb{G}_\diamond$. The following rough heat-kernel estimates will suffice for our purposes:

LEMMA 8.7.

(8.17) $\quad p_n^{\mathbb{G}_o}(\mathbf{y}_0, \mathbf{y}) \leq e^{-c(d)n}$,

(8.18) $\quad p_n^{\mathbb{G}_\diamond}(\mathbf{y}_0, \mathbf{y}) \leq n^{-3/5} + c(d, |\mathbf{y}|) \exp\{-c'(d, |\mathbf{y}|) n^{c(d)}\}$ and

(8.19) $\quad p_n^{G_N}(y_0, y) \leq c e^{-c(d)d(y_0, y)} \mathbf{1}_{n \leq N^3} + c(d)\left(d^{-N} + n^{-3/5}\right) \mathbf{1}_{n > N^3}$.

(We refer to the end of the introduction for our convention on constants.)

PROOF. The estimate (8.17) can be shown by an elementary estimate on the biased random walk $(d(Y_n, \mathbf{y}))_{n \geq 0}$ on $\mathbb{N}$. More generally, (8.17) is a consequence of the non-amenability of $\mathbb{G}_o$, see [**47**], Corollary 12.5, p. 125.

We now prove (8.18). Under $P_{\mathbf{y}_0}^{\mathbb{G}_\diamond}$, the height $|Y|$ of Y is distributed as a random walk on $\mathbb{N}$ starting from $|y_0|$ with transition probabilities $w_{k,k+1} = \frac{1}{d+1}$, $w_{k,k-1} = \frac{d}{d+1}$ for $k \geq 1$ and reflecting barrier at 0. We set for $n \geq 1$,

(8.20) $$L = \left\lfloor \frac{3}{5 \log d} \log n \right\rfloor + 1,$$

and define the stopping time S as the first time when Y reaches the level $|\mathbf{y}| + L$:

$$S = \inf\{n \geq 0 : |Y_n| \geq |\mathbf{y}| + L\}.$$

Then we have

$$p_n^{\mathbb{G}_\diamond}(\mathbf{y}_0, \mathbf{y}) \leq P_{\mathbf{y}_0}^{\mathbb{G}_\diamond}[S \leq n, Y_n = \mathbf{y}] + P_{|\mathbf{y}_0|}^{\mathbb{G}_\diamond}[S > n], \text{ for } n \geq 0.$$

8. Examples

Observe that the second probability on the right-hand side can only increase if we replace $|\mathbf{y}_0|$ by 0. We now apply the simple Markov property and this last observation at integer multiples of the time $|\mathbf{y}|+L$ to the second probability and the strong Markov property at time S to the first probability on the right-hand side and obtain

$$(8.21) \quad p_n^{\mathbb{G}_\diamond}(\mathbf{y}_0, \mathbf{y}) \leq E_{\mathbf{y}_0}^{\mathbb{G}_\diamond}\big[S \leq n, P_{Y_S}^{\mathbb{G}_\diamond}[Y_m = \mathbf{y}]\big|_{m=n-S}\big] \\ + P_0^{\mathbb{G}_\diamond}[S > |\mathbf{y}| + L]^{\lceil \frac{n}{|\mathbf{y}|+L}\rceil}.$$

The second probability on the right-hand side is equal to $1 - (d+1)^{-(|\mathbf{y}|+L)}$. In order to bound the expectation, note that by definition of S, there are d^L descendants $\mathbf{y}'$ of Y_S at the same height as $\mathbf{y}$, and the $P_{Y_S}^{\mathbb{G}_\diamond}$-probability that Y_m equals $\mathbf{y}'$ is the same for all such $\mathbf{y}'$. Hence, the expectation on the right-hand side of (8.21) is bounded by d^{-L}. We have hence shown that

$$p_n^{\mathbb{G}_\diamond}(\mathbf{y}_0, \mathbf{y}) \leq \left(\frac{1}{d}\right)^L + \left(1 - \left(\frac{1}{d+1}\right)^{|\mathbf{y}|+L}\right)^{\lceil \frac{n}{|\mathbf{y}|+L}\rceil}.$$

Substituting the definition of L from (8.20) and using that $\frac{\log(d+1)}{\log d} \leq \frac{\log 3}{\log 2} < \frac{5}{3}$ for the second term, one finds (8.18).

We now come to (8.19) and first treat the case $n \leq N^3$. By uniform boundedness and reversibility of the measure $y \mapsto w_y$, we have $p_n^{G_N}(y_0, y) \leq c p_n^{G_N}(y, y_0)$, so we can freely exchange y_0 and y in our estimates. In particular, we can assume that $d(y_0, o) \leq d(y, o)$. Now we denote by y_1 the first vertex at which the shortest path from y_0 to o meets the shortest path from y to o. Then any path from y_0 to y must pass through y_1. From the strong Markov property applied at time H_{y_1}, it follows that

$$(8.22) \quad p_n^{G_N}(y_0, y) = E_{y_0}^{G_N}\big[\{H_{y_1} \leq n\}, P_{H_{y_1}}^{G_N}[Y_k = y]\big|_{k=n-H_{y_1}}\big].$$

The $P_{H_{y_1}}^{G_N}$-probability on the right-hand side remains unchanged if y is replaced by any of the $d^{d(y_1,y)}$ descendants y' of y_1 at the same height as y. Moreover, the assumption $d(y_0, o) \leq d(y, o)$ implies that $d(y_1, y) \geq d(y_1, y_0)$, hence $2d(y_1, y) \geq d(y_0, y)$. In particular, there are at least $d^{d(y_0, y)/2}$ different vertices y' for which $P_{H_{y_1}}^{G_N}[Y_k = y] = P_{H_{y_1}}^{G_N}[Y_k = y']$. By (8.22), this proves the estimate (8.19) for $n \leq N^3$. We now treat the case $n > N^3$. The argument used to prove (8.18) with $(|y|+L) \wedge N$ playing the role of $|y|+L$ yields

$$(8.23) \quad p_n^{G_N}(y_0, y) \leq c(d, |y|)\big(d^{-N} \vee n^{-3/5} + e^{-c(d,|y|)n^{c(d)}}\big).$$

The assumption $n > N^3$ will now allow us to remove the dependence on $|y|$ of the right-hand side. By applying the strong Markov property at the entrance time $H_{\partial B(o, N-1)}$ of the random walk into the set $\partial B(o, N-$

154   4. Random walks on cylinders and random interlacements

1) of leaves of G_N, we have

$$p_n^{G_N}(y_0, y) \leq P_{y_0}^{G_N}[H_{\partial B(o,N-1)} > N^3/2]$$
$$+ \sup_{y': |y'|=0} \sup_{n-N^3/2 \leq k \leq n} p_k^{G_N}(y', y), \text{ for } n > N^3.$$

Applying reversibility to exchange y' and y, then (8.23) to the second term, we infer that

(8.24) $\quad p_n^{G_N}(y_0, y) \leq P_{y_0}^{G_N}[H_{\partial B(o,N-1)} > N^3/2]$
$$+ c(d)\big(d^{-N} + n^{-3/5}\big), \text{ for } n > N^3,$$

where we have used that $e^{-c(d)n^{c(d)}} \leq c(d)n^{-2/3}$. In order to bound the first term on the right-hand side, we apply the Markov property at integer multiples of $10N$ and obtain

(8.25) $\quad P_{y_0}^{G_N}[H_{\partial B(o,N-1)} > N^3/2] \leq \sup_{y \in G_N} P_y^{G_N}[H_{\partial B(o,N-1)} > 10N]^{cN^2}.$

Note that the random walk on $\mathbb{G}_o \supset G_N$, started at any vertex y in $G_N = B(o, N)$, must hit $\partial B(o, N-1)$ before exiting $B(y, 2N)$. Applying this observation to the probability on the right-hand side of (8.25), we deduce with (8.24) that

$$p_n^{G_N}(y_0, y) \leq P_o^{\mathbb{G}_o}[T_{B(o,2N)} > 10N]^{cN^2} + c(d)\big(d^{-N} + n^{-3/5}\big), \text{ for } n > N^3.$$

The probability on the right-hand side is bounded by the probability that a random walk on $\mathbb{Z}$ with transition probabilities $p_{z,z+1} = d/(d+1)$ and $p_{z,z-1} = 1/(d+1)$ starting at 0 is at a site in $(-\infty, 2N]$ after $10N$ steps. From the law of large numbers applied to the iid increments with expectation $(d-1)/(d+1) \geq 1/3$ of such a random walk, it follows that this probability is bounded from above by $1 - c < 1$ for $N \geq c'$, hence bounded by $1 - c'' < 1$ for all N (by taking $1 - c'' = (1-c) \vee \max\{P_o^{\mathbb{G}_o}[T_{B(o,2N)} > 10N] : N < c'\}$). It follows that

$$p_n^{G_N}(y_0, y) \leq e^{-c(d)N^2} + c(d)\big(d^{-N} + n^{-3/5}\big)$$
$$\leq c(d)\big(d^{-N} + n^{-3/5}\big), \text{ for } n > N^3.$$

This completes the proof of (8.19) and of Lemma 8.7. $\square$

We now consider vertices y_m in G_N that remain at a height that is either of order N or constant. This gives rise to the two different transient limit models $\mathbb{G}_o \times \mathbb{Z}$ and $\mathbb{G}_\diamond \times \mathbb{Z}$.

THEOREM 8.8. $(d \geq 2)$ *Consider vertices $x_{m,N}$, $1 \leq m \leq M$, in $G_N \times \mathbb{Z}$ satisfying A3 and A4 and assume that for some number $0 \leq M' \leq M$ and some $\delta \in (0,1)$,*

(8.26) $\quad \liminf_N |y_{m,N}|/N > \delta, \text{ for } 1 \leq m \leq M', \text{ and}$

(8.27) $\quad |y_{m,N}| \text{ is constant for } M' < m \leq M \text{ and large } N.$

8. Examples

Then the conclusion of Theorem 1.1 holds with $\mathbb{G}_m = \mathbb{G}_o$ for $1 \leq m \leq M'$, $\mathbb{G}_m = \mathbb{G}_\diamond$ for $M' < m \leq M$ and $\beta = 1$.

PROOF. Once more, we check A1-A10 and (1.7) and apply Theorem 1.1. It is immediate to check A1. For the estimate A2, the degree of the root of the tree does not play a role, as can readily be seen from the definition (2.9) of λ_N. We can hence change the degree of the root from $d+1$ to d and apply the estimate from Aldous and Fill in [3], Chapter 5, p. 26, equation (59). Combined with Lemma 8.1 relating the discrete- and continuous time spectral gaps, this shows that $c(d)|G_N|^{-1} \leq \lambda_N$. In particular, A2 holds. We are assuming A3 and A4 in the statement. For A5, we define

$$r_N = \frac{1}{4^M 10}\Big(\min_{1 \leq m < m' \leq M} d(x_m, x_{m'}) \wedge \delta N\Big), \text{ as well as}$$

$o_m = o$ for $1 \leq m \leq M'$ and $o_m = (|y_m|; \mathbf{1})$, for $M' < m \leq M$,

where $\mathbf{1}$ denotes the infinite sequence of ones. Then for $1 \leq m \leq M'$, the ball $B(y_m, r_N)$ does not contain any leaves of G_N for large N, so there is an isomorphism ϕ_m mapping $B(y_m, r_N)$ to $B(o, r_N) \subset \mathbb{G}_o$. For $M' < m \leq M$, note that assumption (8.27) and the choice of r_N imply that for large N, all vertices in the ball $B(y_m, r_N)$ have a common ancestor $y_* \in G_N \setminus (B(y_m, r_N) \cup \{o\})$ (we can define y_* as the first vertex not belonging to $B(y_m, r_N)$ on the shortest path from y_m to o). We now associate a label $l(y)$ in $\{1, \ldots, d\}$ to all descendants y of y_* in the following manner: We label the d children of y_* by $1, \ldots, d$ such that the vertex belonging to the shortest path from y_* to y_m is labelled 1. We then do the following for any descendant y of y_*: If one of the children of y belongs to the shortest path from y_* to y_m, we associate the label 1 to this child and associate the labels $2, \ldots, d$ to the remaining $d-1$ children in an arbitrary fashion. If none of the children of y belong to the shortest path from y_* to y_m, we label the d children of y by $1, \ldots, d$ in an arbitrary fashion. Having labelled all descendants of y in this way, we define for any descendant y of y_* the finite sequence $s(y)$ by $l(y), l(y_1), \ldots, l(y_{d(y,y_*)-1})$, where $(y, y_1, \ldots, y_{d(y,y_*)-1}, y_*)$ is the shortest path from y to y_*. Then the function ϕ_m from $B(y_m, r_N)$ to $\mathbb{G}_\diamond$, defined by

$$(8.28) \qquad \phi_m(y) = (|y|; s(y), 1, 1, \ldots),$$

is an isomorphism from $B(y_m, r_N)$ into $\mathbb{G}_\diamond$ mapping y_m to $(|y_m|; \mathbf{1})$, as required. Hence, A5 holds. As in the previous examples, we now choose the sets C_m ensuring that the probability of escaping to the complement of a large box from the boundaries of B_m (cf. (5.3)) is large. We define the auxiliary graphs as $\hat{\mathbb{G}}_m = \mathbb{G}_m$. As in the example of the box, we then apply Lemma 3.2 to find the required sets C_m. Applied to the points $y_1, \ldots, y_m$, with $a = \frac{\delta}{4^M 10}N$ and $b = 2$, Lemma 3.2 yields points

$y_1^*, \ldots, y_M^*$, some of which may be identical, and a p between $\frac{\delta}{4M^{10}}N$ and $\frac{\delta}{10}N$ such that for $1 \leq m \leq M$,

(8.29) either $C_m = C_{m'}$ or $C_m \cap C_{m'} = \emptyset$ for $C_m = B(y_m^*, 2p)$,

and such that the balls with the same centers and radius p still cover $\{y_1, \ldots, y_M\}$. Since $r_N \leq p$, we can associate a set C_m to any $B(y_m, r_N)$ such that A6 holds. Concerning A7, note that the definition of r_N immediately implies that $\bar{C}_m$ contains leaves of G_N if and only if $m > M'$ and in this case all vertices in $\bar{C}_m$ have a common ancestor in $G_N \setminus (\bar{C}_m \cup \{o\})$ (one can take the first vertex not belonging to $\bar{C}_m$ on the shortest path from y_m to o). We can hence define the isomorphisms ψ_m from $\bar{C}_m$ into $\hat{\mathbb{G}}_m$ in the same way as we defined the isomorphisms ϕ_m above, so A7 holds. Assumption A8 directly follows from (8.29). We now turn to A9. For $1 \leq m \leq M'$, this assumption is immediate from (8.17). For $M' < m \leq M$, note that the isomorphism ψ_m, defined in the same way as ϕ_m in (8.28), preserves the height of any vertex. In particular, $|\psi_m(y_m)|$ remains constant for large N by (8.27) and the estimate required for A9 follows from (8.18). In order to check A10, we again use Lemma 8.2 and only verify (8.5). Note that for any $1 \leq m \leq M$, the distance between vertices $y_0 \in \partial(C_m^c)$ and $y \in B(y_m, \rho_0)$ is at least $c(\delta, M, \rho_0)N$. With the estimate (8.19) and the bound on λ_N^{-1} shown under A2, we find that the sum in (8.5) is bounded by

$$N^3 cd^{-c(\delta, M, \rho_0)N} + c(d)\Big(|G_N|^{-(1-\epsilon)/2} + \sum_{n=N^3}^{\infty} n^{-3/5-1/2}\Big),$$

which tends to 0 as N tends to infinity for $0 < \epsilon < 1$. We have thus shown that A10 holds. Finally, we check (1.7). To this end, note first that all vertices in $G_{N-1} \subset G_N$ have degree $d+1$ in G_N, and the remaining vertices of G_N (the leaves) have degree 1. Hence,

(8.30) $\quad \dfrac{w(G_N)}{|G_N|} = \dfrac{|G_{N-1}|}{|G_N|}\dfrac{d+1}{2} + \Big(1 - \dfrac{|G_{N-1}|}{|G_N|}\Big)\dfrac{1}{2}.$

Now G_N contains one vertex of depth 0 (the root) and $(d+1)d^{k-1}$ vertices of depth k for $k = 1, \ldots, N$. It follows that $|G_N| = 1 + (d+1)(1+d+\ldots+d^{N-1}) = 1 + \frac{d+1}{d-1}(d^N - 1)$ and that $\lim_N |G_{N-1}|/|G_N| = 1/d$. With (8.30), this yields

$$\lim_N \frac{w(G_N)}{|G_N|} = \frac{d+1}{2d} + \frac{d-1}{2d} = 1.$$

Therefore, (1.7) holds with $\beta = 1$. The result follows by application of Theorem 1.1. $\square$

REMARK 8.9. The last theorem shows in particular that the parameters of the Brownian local times and hence the parameters of the random interlacements appearing in the large N limit do not depend on the degree $d+1$ of the tree. Indeed, we have $\beta = 1$ for any $d \geq 1$.

8. Examples

The above calculation shows that this is an effect of the large number of leaves of G_N. This behavior is in contrast to the example of the Euclidean box treated in Theorem 8.3, where the effect of the boundary on the levels of the appearing random interlacements is negligible.

Bibliography

[1] D.J. Aldous. *Probability Approximations via the Poisson Clumping Heuristic.* Springer-Verlag, 1989.
[2] D.J. Aldous and M. Brown. Inequalities for rare events in time-reversible Markov chains I. *Stochastic Inequalities.* M. Shaked and Y.L. Tong, ed., IMS Lecture Notes in Statistics, volume 22, 1992.
[3] D.J. Aldous and J. Fill. *Reversible Markov chains and random walks on graphs.* http://www.stat.Berkeley.EDU/users/aldous/book.html.
[4] N. Alon, J.H. Spencer, P. Erdős. *The Probabilistic Method.* John Wiley & Sons, New York, 1992
[5] M. Barlow, T. Coulhon, T. Kumagai. Characterization of sub-gaussian heat kernel estimates on strongly recurrent graphs. *Communications on Pure and Applied Mathematics*, Vol. LVIII, 1642-1677, 2005.
[6] I. Benjamini, A.S. Sznitman. Giant component and vacant set for random walk on a discrete torus. *J. Eur. Math. Soc. (JEMS)*, 10(1):133-172, 2008.
[7] K.L. Chung. *A course in probability theory.* Academic Press, San Diego, 1974.
[8] C.S. Chou, P.A. Meyer. Sur la représentation des martingales comme intégrales stochastiques dans les processus ponctuels. In *Séminaire de Probabilités IX*, volume 465 of *Lecture Notes in Mathematics*, 226-236. Springer-Verlag, Berlin, 1975
[9] R.W.R. Darling, J.R. Norris. Differential equation approximations for Markov chains. *Probability Surveys* 5, 37-79, 2008.
[10] A. Dembo, Y. Peres, J. Rosen, O. Zeitouni. Cover times for Brownian motion and random walks in two dimensions. *Ann. Math.*, 160, 433-464, 2004.
[11] A. Dembo, Y. Peres, J. Rosen. How large a disc is covered by a random walk in n steps? *Ann. Probab.* 35, 577-601, 2007.
[12] A. Dembo and A.S. Sznitman. On the disconnection of a discrete cylinder by a random walk. *Probability Theory and Related Fields*, 136(2): 321-340, 2006.
[13] A. Dembo, A.S. Sznitman. A lower bound on the disconnection time of a discrete cylinder, *Progress in Probability, vol. 60, In and Out of Equilibrium 2*, Birkhäuser, Basel, 211-227, 2008.
[14] J.D. Deuschel and A. Pisztora. Surface order large deviations for high-density percolation. *Probability Theory and Related Fields*, 104(4): 467-482, 1996.
[15] R. Durrett. *Probability: Theory and Examples.* (third edition) Brooks/Cole, Belmont, 2005
[16] P. Erdős, P. Révész. Three problems on the random walk in $\mathbb{Z}^d$, *Studia Scientiarum Mathematicarum Hungarica*, 26, 309-320, 1991.
[17] O.D. Jones. Transition probabilities for the simple random walk on the Sierpinski graph. *Stochastic Processes and their Applications.* 61:45-69, 1996.
[18] R.Z. Khaśminskii. On positive solutions of the equation $\mathcal{U} + Vu = 0$. *Theory of Probability and its Applications*, 4:309-318, 1959.
[19] G.F. Lawler. *Intersections of random walks.* Birkhäuser, Basel, 1991
[20] G. Lawler. On the covering time of a disc by a random walk in two dimensions. *Seminar in Stochastic Processes*, 189-208, Birkhäuser, Boston, 1993.

[21] L.H. Loomis and H. Whitney: An inequality related to the isoperimetric inequality. *Bulletin of the American Mathematical Society*, 55: 961-962, 1949
[22] E.W. Montroll. Random walks in multidimensional spaces, especially on periodic lattices. *J. Soc. Industr. Appl. Math.*, 4(4), 1956.
[23] J.R. Norris. *Markov Chains*. Cambridge University Press, New York, 1997.
[24] G. Pólya. Über eine Aufgabe der Wahrscheinlichkeitsrechnung betreffend der Irrfahrt im Straßennetz. *Math. Annalen*, 84, p. 149-160, 1921.
[25] P. Révész. Local time and invariance. *Analytical Methods in Probability Theory*, Lecture Notes in Mathematics, 861:128-145. Springer, Berlin, Heidelberg, 1981.
[26] P. Révész. On the volume of the spheres covered by a random walk, *A tribute to Paul Erdős*, A. Baker, B. Bollobás, A. Hanjal (editors), Cambridge University Press, Cambridge, 341-347, 1990.
[27] P. Révész. Clusters of a random walk on the plane, *The Annals of Probability*, 21(1), 318-328, 1993.
[28] L. Saloff-Coste. *Lectures on finite Markov chains*, volume 1665. Ecole d'Eté de Probabilités de Saint Flour, P. Bernard, ed., Lecture Notes in Mathematics, Springer, Berlin, 1997.
[29] T. Shima. On eigenvalue problems for the random walks on the Sierpinski pre-gaskets. *Japan J. Indust. Appl. Math.*, 8, 127-141, 1991.
[30] V. Sidoravicius and A.S. Sznitman. Percolation for the vacant set of random interlacements. *Communications on Pure and Applied Mathematics*, to appear, also available at http://arxiv.org/abs/0808.3344.
[31] V. Sidoravicius and A.S. Sznitman. Connectivity bounds for the vacant set of random interlacements, preprint available at http://www.math.ethz.ch/u/sznitman/preprints.
[32] P.M. Soardi. *Potential Theory on Infinite Networks*. Springer-Verlag, Berlin, Heidelberg, New York, 1994.
[33] A.S. Sznitman. Slowdown and neutral pockets for a random walk in random environment. *Probability Theory and Related Fields*, 115:287-323, 1999.
[34] A.S. Sznitman. On new examples of ballistic random walks in random environment. *The Annals of Probability*, 31(1):285-322, 2003
[35] A.S. Sznitman. How universal are asymptotics of disconnection times in discrete cylinders? *The Annals of Probability*, to appear.
[36] A.S. Sznitman. Random walks on discrete cylinders and random interlacements, *Probability Theory and Related Fields*, to appear, preprint available at http://arxiv.org/abs/0805.4516.
[37] A.S. Sznitman. Upper bound on the disconnection time of discrete cylinders and random interlacements, *Annals of Probability*, to appear, preprint available at http://www.math.ethz.ch/u/sznitman/preprints.
[38] A.S. Sznitman. On the domination of random walk on a discrete cylinder by random interlacements, preprint available at http://www.math.ethz.ch/u/sznitman/preprints
[39] A.S. Sznitman. Vacant set of random interlacements and percolation, *Annals of Mathematics*, to appear, preprint available at http://arxiv.org/abs/0704.2560.
[40] A.Q. Teixeira. On the uniqueness of the infinite cluster of the vacant set of random interlacements. *The Annals of Applied Probability*, Vol. 19, No. 1, 454-466, 2009.
[41] A.Q. Teixeira. Interlacement percolation on transient weighted graphs, preprint available at http://www.math.ethz.ch/∼teixeira/.
[42] A. Telcs. *The art of random walks*. Lecture Notes in Mathematics, 1885. Springer, Berlin, Heidelberg, 2006.

[43] D. Windisch. On the disconnection of a discrete cylinder by a biased random walk. *The Annals of Applied Probability*, Volume 18, Number 4, 1441-1490, 2008.
[44] D. Windisch. Logarithmic components of the vacant set for random walk on a discrete torus. *Electronic Journal of Probability*, 13, 880-897, 2008.
[45] D. Windisch. Random walk on a discrete torus and random interlacements. *Electronic Communications in Probability*, 13, 140-150, 2008.
[46] D. Windisch. Random walks on discrete cylinders with large bases and random interlacements, *The Annals of Probability*, to appear.
[47] W. Woess. *Random walks on infinite graphs and groups*. Cambridge University Press, Cambridge, 2000.

VDM Verlagsservicegesellschaft mbH

Die VDM Verlagsservicegesellschaft sucht für wissenschaftliche Verlage abgeschlossene und herausragende

Dissertationen, Habilitationen, Diplomarbeiten, Master Theses, Magisterarbeiten usw.

für die kostenlose Publikation als Fachbuch.

Sie verfügen über eine Arbeit, die hohen inhaltlichen und formalen Ansprüchen genügt, und haben Interesse an einer honorarvergüteten Publikation?

Dann senden Sie bitte erste Informationen über sich und Ihre Arbeit per Email an *info@vdm-vsg.de*.

Sie erhalten kurzfristig unser Feedback!

VDM Verlagsservicegesellschaft mbH
Dudweiler Landstr. 99
D - 66123 Saarbrücken
Telefon +49 681 3720 174
Fax   +49 681 3720 1749
www.vdm-vsg.de

Die VDM Verlagsservicegesellschaft mbH vertritt

Printed by Books on Demand GmbH, Norderstedt / Germany